Windpower Principles: their application on the small scale

Experimental aerofoil and sail mills on test rigs

Windpower Principles: their application on the small scale

N.G.CALVERT

Ph.D., F.I.Mech.E.
Department of Mechanical Engineering,
University of Liverpool

Illustrations by F. Cummins

A Halsted Press Book

John Wiley & Sons
New York – Toronto

CHARLES GRIFFIN & COMPANY LIMITED
Registered Office:
Charles Griffin House, Crendon Street, High Wycombe
Bucks, HP13 6LE, England

First published 1979

Published in the U.S.A. and Canada by Halsted Press,
A Division of John Wiley & Sons, Inc., New York

Library of Congress Cataloging in Publication Data
Calvert, N. G.
Windpower principles: their application on the small
scale.
"A Halsted Press book."
Bibliography: p.
Includes index.
1. Wind power.   I. Title.
TJ825.C28   621.4'5   79–19706
ISBN 0–470–26867–0

Text set in 10/12 pt Photon Baskerville, printed and bound
in Great Britain at The Pitman Press, Bath

Contents

Preface

The elemental powers of wind and water had been used for centuries before steam and other engines made small local and intermittent sources of energy uneconomic. The assumption that indefinite supplies of cheap fuel would continue to remain available is no longer valid and it is desirable, at the least, to remember that other sources of energy exist. Water (including tidal) power is well understood but it is applicable only to limited geographical areas. The direct utilisation of solar radiation involves research of the most fundamental character; meanwhile, we can be well assured that so long as the earth remains habitable, then the winds will continue to blow.

As yet we cannot demonstrate large windpower units in daily use but many minds are at work and, when the time is ripe, progress will be made. In the meantime, small-scale installations, designed to suit the location and the duty *can* be a practical proposition. Even these involve meeting quite complex conditions which the book sets out to explain.

Classical accounts of windpower may be found in the proceedings of learned societies, current work in the reports of research organisations, and these are indicated in the references. Inventors such as Flettner and Savonius wrote their own monographs, and comments on their work can be found in the technical periodicals of the time. The windmill on the "Fram" is mentioned in Nansen's reports of polar exploration. Miscellaneous scraps of information may be found in travel books.

The author's search for the origins of the Beaufort Scale of Wind Force began in the older editions of the Dictionary of National Biography. It was sufficiently completed with the help of the National Maritime Museum. Information on sail mills stems from the travels and observations of the author and his wife in the Greek Islands. In this they received the support of the British School at Athens. Numerical results, unless ascribed to others, are generally those observed by the author or by students under his supervision.

An invitation to deliver a public lecture on the subject led to the author devising demonstration models. This lecture evidently aroused interest for more invitations followed and more experiments and illustrations of current field work were devised. The book follows the plan of the lectures in that it aims to be comprehensible to the interested non-specialist and yet not too trivial to interest a practising engineer.

A book such as this cannot encompass details of mechanical design. To attempt to do so could be dangerous if it gave the impression that a single solution could suit all circumstances. Also the many for whom the book is written will build for individual sites and duties and they will adopt such useful material as is at hand. The book aims to give *theoretical* guidance to those who already have sufficient constructional engineering knowledge and practical experience.

The essential knowledge of fluid mechanics is presented in a conceptual rather than an analytic form. From this the properties of an aerofoil are explained and an aerodynamic system of classification is developed. Many types of windmill are considered and particular attention is given to the sail mill, for which the author has a special regard.

Methods of measurement and systematic testing are explained in some detail in order that the potentialities of the windmill as a small alternative energy source and/or as an aid to education (both in the field and in the laboratory) can be appreciated. Guidance is given to the prospective builder as to the amount of energy he can expect, and of certain risks he might run.

Summarising, both the keen amateur with technical ability, and the practising engineer reviewing this field of power production for the first time, will find all aspects explained in a logical and orderly way. It has been written not only with the background of a long engineering experience, but also from many years of personal research and experiment with small wind power plant and its performance.

Liverpool, Oct., 1978   N. G. CALVERT

Dedicated to

Experimenters and writers of the past who left a record of
their observations for others to use; to all those who have in
their various ways helped the author to complete this book;
and most of all to his wife.

1 Background to wind power

Historic

The existence of the sailing ship, as proved by Egyptian records and Cretan seals, indicates that for four thousand years some members of mankind have been experienced in handling quite substantial units of wind power.

The first use of wind to secure rotary motion is not known. There is still scope for scholarship and archaeology. As regards western Europe, certainty comes during the thirteenth century. It was the custom of scribes to decorate their documents with little illuminated pictures of scenes from daily life. Clearly recognisable windmills, complete with people carrying bags of corn, appear at this time. Recent studies by Mr. Stergios Spanakis, the historian of Heraklion, have shown their use on the island of Crete as an adjunct to the fortress installations during the Venetian period. The northern windmills were of the post type which could be turned to face the wind (Fig. 1.1). The Cretan ones were probably monokairos, that is "single weather", or fixed to face the prevailing wind (Fig. 1.2). Both may have started with reed or matting sails interlaced between bars projecting from the sail arms. The northern windmills developed into large and complicated machines. The Cretan corn mills can hardly have developed at all (except for the change to fabric sails) so basic are the structures, which can be seen today. A few were in use only a year or so ago. One marvels that the Cretan can build so much out of so little. It is a matter for wonder that windmills, so far as we know, appeared so suddenly and so complete in places so far apart. The developments which led up to them may never be known.

Historical evidence suggests that the first use of the windmill was for the grinding of corn and in Europe it is probable that this was in the Low Countries since they have very little water power. Its early application in this same area to land drainage was stimulated by desperate need. The Netherlands suffered dreadful inundations

Fig. 1.1   Post mill

during the fourteenth century, when the Zuider Zee was formed, and there was also a specific disaster in 1421. In 1608 windmill pumping was in hand on a really large scale to eliminate large inland lakes which were threatening to engulf the cities (Fig. 1.3). But this was not all. A whole industrial complex based on wind power arose in the Netherlands. Windmills were, for example, also used for sawing wood, extracting oil, grinding stone and in paper making.

A different kind of pumping duty came later, in other lands, that of lifting much smaller amounts of water, for domestic or stock watering purposes, through much bigger heights. It could come from a deep well and be delivered to an elevated tank. This kind of machine needed a large initial force to overcome the inertia and the pressure of the water. A new kind of wind machine generally known as the "American" windpump appeared. Its characteristic feature was the larger number of blades on the wind wheel. While the traditional mill generally used four, exceptionally two, five, or six, the American windpump (Fig. 1.4), in its earliest form with narrow wooden blades,

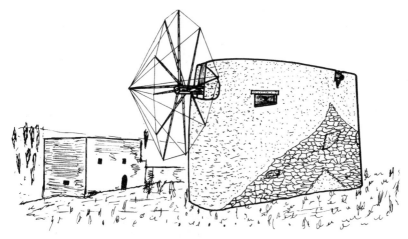

Fig. 1.2 Monokairos (unidirectional) mill

is recorded as having up to 160 blades in a 5 m (16 ft) diameter. Subsequently these machines were factory built of iron with perhaps 24 blades arranged around the circumference. Size tended to optimise at about 2.5 m (8 ft) diameter, although much larger ones have been built. In this form they are well known, being reported in their thousands on the plains of North and South Africa, Australia and America. They were not unknown in Britain until almost universal public water supplies rendered most of them redundant. Their basic feature is that they start readily at low wind speeds and so work for many hours per year.

The nineteen-twenties were a time of hope and advance; aeronautics under the pressure of war had advanced apace. The science of fluid mechanics was about to be born. This new knowledge was turned directly toward the perfection of a special windmill, not for generating power down on earth but for supplementary drives on aircraft, in particular for powering radio and for fuel pumps. These little windmills had to aim at aerodynamic and mechanical perfection for their power came indirectly from the main engines and also weight was vital.

Interesting use of windpower during the 1939–45 war included windmill-powered retrieval winches on target-towing aircraft and a booster slung below the low-level, ship-flown barrage balloons. A shrouded free-running rotor had four blades the outer half of which,

Fig. 1.3   Tower mill

at fine pitch, worked in the airstream and the inner half, at a coarse pitch, formed a blower impeller delivering air at nearly double the free-air pressure to the ballonet of the balloon (Fig. 1.5a).

The follow-up from such applications has been the production of what are often called propeller windmills, that is, machines with blades shaped something like an aeroplane propeller which could run fast enough to drive a dynamo without the use of gears (Fig. 1.5b). These small machines are one of the great success stories of the windmill and today their descendants, with the advantage of another fifty years development in technology, are still being built. Like many good things they are expensive and it is only in special circumstances that they can be financially attractive when compared to public

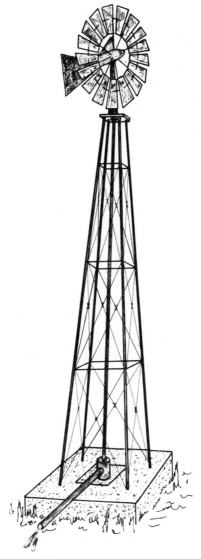

Fig. 1.4   American mill

supplies or internal combustion engines. They have the advantage of the great intrinsic strength of small things. Difficulties come thick and fast when the size is much increased. You must note that the propeller

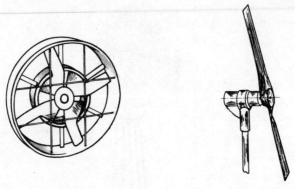

Fig. 1.5   (A) Barrage balloon booster   (B) Propeller windmill

and the high speed windmill are not interchangeable although at first glance they may look much alike.

People in the Netherlands, dismayed by the rapid disappearance of what was the principal vertical feature of their landscape and almost the symbol of their country, the windmill, formed a society for their preservation. Many thousands of windmills had been destroyed, superseded by central station power or by the internal combustion engine. Attempts were made to find ways of improving the traditional mills without much altering their appearance. Aerodynamic knowledge was applied in such a way as to improve the light-weather performance and so increase the hours of work per year. There was no attempt to increase the maximum power for this would have meant complete reconstruction of a device which had been developing over hundreds of years. The method adopted was to shape the leading edge of the sails with sheet metal, making this edge a little more like that of an aeroplane wing. It was very successful and must have kept many mills in work while the preservation movement was gathering strength. The name of A. J. Dekker was associated with this and much of his work can still be seen.

At the same period the people of Crete were emerging into a new circumstance and a local blacksmith of Lasithi adapted a miniature version of the Aegean corn mill to irrigation pumping. With towers first of stone and then in timber and finally in angle iron, the irrigation mills multiplied until there arose a sight without parallel in the world. Swords into ploughshares had its counterpart as artillery cartridge cases were re-shaped to form pump barrels. In due course

Lasithi, the high plateau below the mountain where mythology holds that Zeus was born, became the plateau of ten thousand windmills (Fig. 1.6).

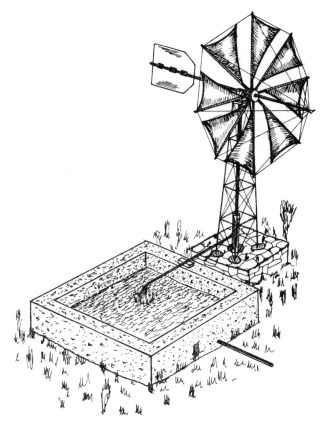

Fig. 1.6 Cretan wind pump

Wind and water power

The sun pours energy down on the earth and, at very much the same rate, the earth radiates it back into space. If it were not so balanced our world would long ago have been either frozen solid or melted away.

Any particular bit of energy, however, does not necessarily leave immediately or from the same place. Some indeed, caught up in organic nature, may have resided here for millions of years as coal or as oil. The tropical lands receive an excess of energy and the temperate lands a deficiency. The transport of this excess is a function of the atmosphere and of the oceans. Changes of water vapour content take a prominent part in this transport of energy. Evaporation and condensation, expansion and contraction, the giving and receiving of heat, add up to a mighty steam engine, which, combining its action with the rotation of the earth, gives us the weather which we know. We might use a tiny bit of the energy of the weather, transiently, for our own purposes, but this is utterly insignificant compared with the mighty flow around us and we do not hold it for long. Very soon it must join in the inexorable process of degradation into low-temperature heat and await its turn to be re-radiated back into space.

Water power is presented to us in a potential form, by being collected in, say, a mountain lake some height above the sea, but it is made manifest to man by its velocity as it runs down as a river or a stream to the sea. Perhaps it is reasonable that such energy should have been first extracted in the velocity form.

Windpower, by definition, is manifested by motion of the atmosphere. Pressure changes take place as well, but these, while they are not of direct importance for the extraction of energy, are of vital importance in the study of the weather and are thus of interest to the windpower engineer.

Power and energy

One of the engineering institutions gives as one of its objectives "The direction of the great sources of energy in nature to the use and convenience of man". Although there is now less confidence in the overall benefits of material progress, it is an objective that can still be accepted, albeit with qualifications. In that such energy has been delivered abundantly, the institution concerned may be said to have succeeded in this objective.

Energy grades

Energy is graded by engineers. A high-grade energy is capable of complete and spontaneous degradation into a lower grade. The reverse process, the upgrading of energy, is only exceptionally and

partially possible. Electricity is the highest form of energy and heat is
the lowest. From the above quoted objective Prometheus may be
considered the first civil engineer, but he stopped short at the very
lowest grade.

As an illustration of changes in energy grade, we all know that the
brakes of a car get hot when they are slowing the vehicle down. The
mechanical energy of the car will appear as heat in the brake drums.
We cannot reverse this process. Cooling the hot brakes does not set
the car in motion again.

Mechanical energy is of only a slightly lower grade than electricity.
When mechanical energy is produced by thermal power, only about
one-third, at the best, leaves the power plant in a high grade form.
The rest is normally wasted, often in a damaging and polluting form.

It is very easy from a position of affluence to point to the dangers
inherent in the access to enormous sources of energy, but quite a
different matter voluntarily to forego it. Perhaps Lord Acton's terri-
ble dictum, "power tends to corrupt," applies to mechanical as well
as political power. There is no limit to our demand for mechanical
slaves. For thousands of years man has sought energy other than his
own to help with transport and for hundreds of years to grind his
corn. Now there are mechanical food mixers, garbage pulverisers,
lawn mowers and even toothbrushes, and all the time the machines
are making more machines. When it comes to personal transport our
demands are really frightening.

The mechanical power which a man can exert can be conveniently
demonstrated by means of a hand-driven dynamo. Such machines
exist, particularly developed for working wireless sets in lifeboats. It
can be an enlightening experience to work one. A strong man soon
gets tired of developing fifty watts. A vacuum cleaner needs five hundred and a small car has the equivalent of twenty thousand watts at its
engine shaft and, at best, four times as much is thrown away in friction and in the exhaust and the radiator. We know if we think about
these things that this pattern of life, so very recent in man's history,
cannot go on.

As the energy stress builds up the question will again be asked,
"What about the wind?" Wind energy is real enough and it is of a
high-grade form, but only a few are in a position to use it and then
on a scale quite insignificant compared with what we have begun to
expect as normal. Small though it may be compared with recent demand it could be very useful indeed to someone who has nothing else

... and you can always build a little windmill to help with your hobbies and to educate your children in the realities of power.

Windpower, the ever present manifestation of solar energy and terrestrial thermodynamics, is known to all, but perhaps it is the yacht (or dinghy) helmsman who has most appreciation of its magnitude and usefulness. The way in which a boat pulls away when its head sail fills really puts its auxiliary engine into perspective. An 11 m (36 ft) yacht probably optimises most of the variables of wind energy. A boat has hardly any resistance when its speed is low and so the lightest airs are of value. At the other extreme it can yield to the stress of the weather and so avoid damage in a sudden squall. Air flow over water is less impeded than that over land and it is therefore more steady. Also, its general direction is of only secondary importance. By cunning interaction of wind and water forces, every wind can be put to use. Up to a point, as speed increases, the resistance of the boat keeps step with the enhanced force of the wind, but only up to a point. There comes a time when the boat can go no faster and then the sails must be reefed before disaster overtakes.

Things are altogether more difficult over the land. Bushes and trees however are efficient absorbers of wind energy. Man-made structures are not so effective in this respect. The violent turbulence in an urban area gives a misleading impression of available energy. Eddies around buildings which can lift a man off his feet may signify practically nothing as regards a steady production of power. The author knows because he has erected small windmills in such positions. Also, land-based machines have another considerable but unavoidable characteristic. This is "static friction". It takes a bigger force to get something started than to keep it moving. A minimum force is required to rotate a windmill, or move a land yacht. Forces below this minimum are useless. A gentle on-shore wind which comes up with the dawn and brings the fleet into harbour will not set the windmills turning.

It is, in fact, extremely easy to develop windpower on a human scale, something which can often give you a hundred or two hundred watts of mechanical power to work a grindstone, a drilling machine or a water pump, that is when the wind is suitable and you are at hand to control it. Try, however, to step it up to be of use in a regimented industrial society and difficulties come thick and fast. The low density of the air and the great range of its velocity pose daunting problems.

2 Mechanical considerations

"Let Newton be"

Practical engineering has, until recently, tended to precede the scientific knowledge on which it is based. Steam railways had almost encircled the globe before the laws of thermodynamics were formulated and it was not until aeroplanes began to hop off the ground that former "proofs" of the impossibility of practical mechanical flight began to be questioned. Empirical knowledge of the behaviour of air and water had been accumulating for thousands of years, yet only one hundred years ago Osborne Reynolds (mathematician and engineer) was able to say, "The slightest problem of fluid motion is as yet unsolved".

Newton's laws of motion opened the way to the understanding of much of the mechanical behaviour of the solid world but they had less success when they were applied to fluid problems. It was not for want of trying. Elegant mathematical solutions to fluid problems were created in plenty but they had one thing in common. When compared with experimental observations they were always wrong. So universal was this discrepancy that it has passed into the language as the "Paradox of D'Alembert". The reason for such failure was of course that significant variables had been omitted from the assumptions made before the calculations were begun. This followed from lack of experimental evidence as to what the fluid really did.

In British engineering history pride of place often goes to the self-taught men of the Industrial Revolution: Brindley, Telford, Stephenson and the rest. Craftsmen themselves but with great intelligence and insight, they had little reason to look up to the scientific establishment of the time. They accordingly built their own monuments and they moulded the professional institutions in their own image so that their influence is with us today. More enduring than the monumental works of any of these are the ideas based on experimental observation of the generation of engineer-scientists who followed them. Resounding names: James and William Thompson (Lord Kelvin), Rankine, Osborne Reynolds, William and Edmund Froude, and Hele-Shaw. Unlike the founding fathers these people came from privileged and cultured homes, not uncommonly they were the sons of clergymen, and they graduated in mathematics. Mostly they became professors of engineering in the provincial universities and their common feature was that when faced with a problem they would dig down to its very foundation either in the field or in the laboratory, and they were not ashamed to dirty their hands.

The science of fluid mechanics is recent, much of it having been established in the last fifty years. Although it is capable of mathematical treatment its basis is experimental and an appreciation of its concepts can do much to extend awareness and enjoyment of the natural world. Fluid mechanics knows of no artificial distinction between the natural and the man-made, between organic and inorganic. Its concepts can be seen at work in the flight of birds and of aeroplanes and indeed corresponding phenomena have been established in such diverse situations as the flow of air over a mountain ridge and the flow of blood through a human heart. The author believes that as an experimental and conceptual study fluid mechanics should form a part of a liberal education. Along with any other specialised study a specialised terminology has developed which is only a shorthand way for the concise expression of comprehensive ideas. Because of the way it developed, the terminology is that of the aeroplane but the application is general.

Modes of fluid flow

Early investigation of fluid flow experienced anomalous results. Sometimes a fluid resistance would vary linearly as the speed. Sometimes it would be as the square of the speed. It was Osborne Reynolds who pointed out, purely from experiment, that different modes of flow were possible. One was dominated by viscosity, the other by inertia. These modes are now known as laminar and turbulent flow, respectively. Laminar flow is encouraged by low velocities, small sizes and high viscosity, turbulent flow by the converse. It is common experience that a viscous liquid, treacle, or heavy oil, loses viscosity and runs more freely when it is heated. It may come as a surprise to know that a gas such as air behaves in the opposite way. It is least viscous when cold and gets progressively more viscous as it is heated.

Viscosity of air or of water seems of little consequence under ordinary conditions yet it is the viscosity of these fluids that completely dominates the flow pattern and it does this through the mechanism of the "boundary layer". Failure to take this into account is at the root of the Paradox of D'Alembert.

The boundary layer

When a fluid flows over a solid surface the layer of the fluid in contact with the surface is at rest, relative to the surface, but that a little way out is apparently uninfluenced by the proximity of the surface. There is thus a layer of fluid in transition subject to significant shearing action. This is the boundary layer. More careful study reveals two principal regions within the boundary layer. The inmost layer, next to the surface, is dominated by viscosity and so it is laminar. This is now called the laminar sub-layer. Outside the laminar sub-layer is the turbulent boundary layer. The boundary layer can be easily observed along the side of a moving ship when it is in comparatively calm water. One-hundred feet aft from the bow the thickness of the boundary layer is about two feet.

It is enlightening to follow the progress of the boundary layer from its inception until it is finally shed as a wake.

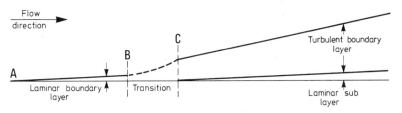

Fig. 2.1   Growth of the boundary layer

Separation

At A (Fig. 2.1), where the flow meets the surface, the boundary layer begins to form. Here it is thin and laminar. The boundary layer is a region where energy is lost and it is always growing in thickness since a slow moving layer retards the next one above it. Soon the laminar boundary layer will have grown so thick that it must become turbulent to survive. This it does at B, the transition point. The turbulent boundary layer now becomes established and begins to grow. (It has also acquired a laminar sub-layer.) In due course the turbulent layer has grown too big and has too little energy to survive. It now peels away, rolls up into a vortex and is shed astern as a wake; this takes place after point C (and at the separation point shown in Fig. 2.5). The control of the point of separation is the control of fluid motion and the forming of a surface which can delay separation right to its end is the proper use of the term streamlining. Fluid friction involves the excitations of the molecules as ordered motion breaks down and disorderly or molecular motion takes its place. This is manifested by a rise in temperature and the problem becomes one of thermodynamics.

Lift and drag

Of importance for our purposes are the forces on a surface which is held inclined to the stream of fluid (Fig. 2.2). The (generally small) angle of the surface to the fluid flow, a, is the angle of attack. The force which the surface experiences perpendicular to the stream is the

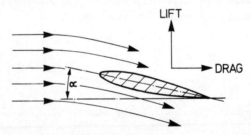

Fig. 2.2   Lift and drag

lift and that which is parallel to it is the drag. Roughly, lift is the force needed to bend the flow. "Lift" in this sense is not necessarily vertical, it is just perpendicular to the stream. It is as correct to talk about lift on a rudder, or a windmill sail, as on a wing. Both lift and drag are strongly influenced by the angle of attack. These concepts are instinctively familiar (but undefined) in daily life. A gentleman wearing a hat who bends his head towards the wind is doing so to reduce the angle of attack and with it both the lift and the drag.

If the flow over our inclined surface is made visible, say with filaments of smoke, it is often seen that the flow bends to follow the underside of the surface, but it does not bend to follow the upper side. This is an example of flow separation. Instead of bending to follow the upper surface it goes straight on leaving a "dead" region (in fact a violently eddying one) between it and the surface. In such a case only the bottom of the surface is producing lift. If separation on the top can be avoided, or even delayed, then the lift is greatly augmented by a suction effect on the top, adding to the pressure effect on the bottom.

All these effects can be seen at work in the natural world. A duck's foot is a clear example of a drag force used to produce thrust when the bird is in the water, while a seagull effortlessly following in the air wake of an Isle of Man steamer can be seen to adjust its angle of attack to suit the circumstances.

The sail is compliant to the fluid forces acting on it and can take up a shape which delays separation. Thus it can produce a bigger lift than can a flat plate. The Aegean millwrights appreciated this but it escaped the attention of the north-west European millwrights until some fifty years ago. Then it was too late. Even so, the sweeps of the western mills show a degree of knowledge which, crudely and incompletely perhaps, anticipates the principles of aerodynamics. The windmill bears some resemblance to a propeller in reverse. Its action could not have been predicted theoretically until Newton's laws were known and the flow of the air made visible. There is no evidence of such knowledge in the early water-mills which appear to have anticipated the windmill by centuries. The invention of the windmill must rank with the invention of the wheel as one of the great achievements of the human intellect. Whilst the idea may have come from experience with boats, the structure of the western windmill is a technology all of its own, bearing little relationship to the methods of the sea. For instance, when sail-cloths are used in the north west, they are not rigged from a spar but stretched over a lattice frame.

The windmills of the Aegean are different and their date of origin is not known. They now use triangular fabric sails rigged in a way which bears a close resemblance to the rig of the local ships. The concepts of spar and stay, of sail and sheet, which have changed but little since the ancient world, are clearly exemplified in the thousands of windmills which to this day are at work in Crete and the Middle East.

The aerofoil

Aeronautical research, mathematical and experimental combined, has produced a great family of shapes, all of them beautiful, called aerofoils. Characteristically they have a blunt nose and a finely tapering tail (Fig. 2.3 a, b, c), and they can either be symmetrical or

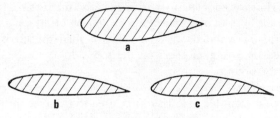

Fig. 2.3   Low-speed aerofoils

otherwise. Their property is that a flow can often follow their curved surfaces without separation (Fig. 2.4). The effect is that they can develop a lift many times, up to fifty times, greater than their drag. The pull on the suction side is generally bigger than the push on the pressure side.

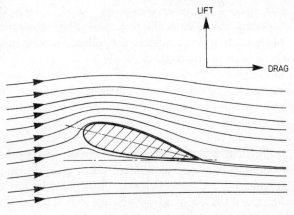

Fig. 2.4   Aerofoil lift

The stall

If the angle of attack becomes too great the flow may suddenly leave the suction side (Fig. 2.5) with a dramatic reduction in lift and in-

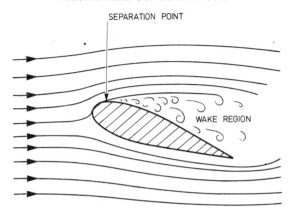

SEPARATION POINT

WAKE REGION

Fig. 2.5   Aerofoil stalled

crease in drag. This is the "stall", a situation which can be catastrophic in mechanical flight, but is part of the instinctive control used by a bird when it comes in to land.

The aerofoil made mechanical flight possible and the concept is inherent in the blades of all jet and turbine engines as well as in the keel and rudder designs of sailing yachts.

Traditional windmills utilise the lift generated on the pressure side only and embody no concept related to the suction of an aerofoil. The compliant sail of either the ship or the Aegean windmill is certainly much nearer to the aerofoil concept.

The angle of attack of a windmill sail may seem to be very large, about 85° when the mill is at rest, but it is very different when the mill is in motion. The apparent, or relative, wind (as it would be to an observer riding on the tip of a windmill sail) is simply shown by a velocity diagram (Fig. 2.6). The diagram (explained in detail below) also shows that the speed of the sail tip may be many times greater than that of the wind itself. Indeed the ratio of tip speed to wind speed is a convenient figure for the classification of windmills. This varies from less than one for an Aegean mill to between two and three for a traditional mill and to between six and ten, or even more, for an aerofoil machine.

The invention of the aerofoil might suggest a windmill such as had not turned before. Indeed it did and some machines are now built on this principle. However, in a windmill context, aerodynamic perfection is not always the most important criterion of excellence.

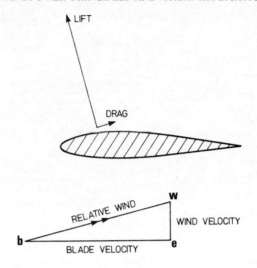

Fig. 2.6   Velocity diagram (symmetrical aerofoil)

Relative velocities

Speed is a concept relating distance travelled and time taken. Velocity is an extension of the concept to include direction as well as speed. The adding and subtracting of velocities is most easily done graphically and is an aid to understanding in many problems relating to fluid motion. The problem is familiar to the sailing boat navigator who, knowing the set of the tide, the direction of the wind, the lateral drift of his boat and its forward speed through the water, needs to know the direction in which he must steer to reach his desired haven.

A similar but less complex problem faces the windmill builder who wishes to find the angle which an aerofoil blade must make to the plane of rotation to achieve a desired blade speed in a given wind. This is easily done with the aid of small scale drawing which can be done in three logical steps. This is illustrated in Fig. 2.6.

Let "e" be a fixed (or earth) point.

(1) Draw a line ew from e whose length represents to some scale the speed of the wind relative to earth and whose direction corresponds to the wind direction.

(2) Draw a line eb from e whose length corresponds to the linear speed of the blade, again relative to the earth, and whose direction corresponds to the blade direction.

(3) The closing line bw now corresponds to the velocity of the wind relative to the blade. This would be the apparent wind to an observer riding on the windmill blade.

The angle of attack is measured from the relative velocity line to the chord of the aerofoil.

Influence of height

The earth, in common with any other surface, exhibits the phenomenon of a boundary layer, that is, the wind velocity will fall as the surface is approached. However the largest windmills which we can imagine are likely to be relatively very low indeed in this boundary layer. Height thus assumes considerable importance, for it is in the lowest layers that the velocity changes are greatest.

For real and steady power, even in the range, say, of twenty to one hundred watts, there is little hope of success unless the windmill is higher than neighbouring obstructions. A machine built in a suburban garden among the fruit trees and rose bushes is unlikely to be more than a decorative mobile. The importance attached to height can be seen in many of the cities of the Netherlands where mill towers are far higher than the neighbouring houses. However, assuming that the site is clear and open, the question arises as to the benefit that may arise from greater height than that necessary for the sails to clear the ground and not be a hazard to passers-by.

The exact way in which velocity varies in the lower levels of a boundary layer is still the subject for recondite research in academic wind-tunnels. Such precision as may be there attained is unthinkable in the lower levels of the atmosphere; nevertheless, it is such basic knowledge which must be used to suggest a rational form for limited empirically based relationships.

Authorities agree that the ratio of the wind speeds V_1 and V_2 between two heights H_1 and H_2 is given by an equation of the form

$$\frac{V_1}{V_2} = \left(\frac{H_1}{H_2}\right)^n$$

where the value of index n depends on the location of the site and the roughness of the ground below it. Values of n vary from about 0.14 over the sea to about 0.34 over rough inland country. These equations put into numerate form the fact that a rough surface will have more effect on the wind than a smooth one.

The real power advantage could be greater than that indicated by increase of velocity, for the higher wind speeds would mean more hours of operation per year, in that the high mill would be working when the low mill was unable to overcome its static friction. The author has seen this in Crete. Small local mills, tucked away among olive trees were becalmed, but a tall American mill was working steadily.

The effect which a practical change of height might have on a small windmill is illustrated in Fig. 2.7. The index of the equation has been assumed to be 0.25. The two curves show the way in which the wind speed and the wind power can be expected to increase as the height is increased.

A hypothetical small windmill can be taken as an example. Initially this windmill is mounted with its axis at a height of 4 m (13 ft) above ground level. (This may be taken as a prudent height in case of someone walking underneath.) The question arises as to what advantage could follow from an increase in height to say 10 m (33 ft). The graph of Fig. 2.7 shows that the wind speed at 10 m (33 ft) can be expected to be 26 per cent greater than at 4 m (13 ft). From the cube law the power increase is greater by a factor of 1.26^3. This is the equal of 2. Thus this increase of height should lead to double the power. A similar calculation shows that an increase of height to 17.3 m (57 ft) could treble it.

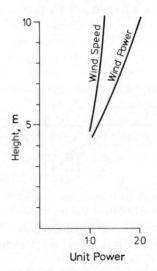

Fig. 2.7 Relation of power to height

But in wind power, as in everything else, any apparent advantage carries its penalties—in this case cost and vulnerability. Most engineering structures can be priced by weight. Assuming that this is so, then the tower of the higher mill, of the first two compared above, being two-and-a-half times as high as the lower, might well cost about fifteen times as much. It might be cheaper and safer to get the extra output by placing two low mills side by side, rather than use the one high one. (Note that this cost ratio refers to the tower only, and not to the machine on top.)

Noise

The transfer of work between a flowing stream and a moving blade is of its essence a discontinuous one. Pressure in the fluid rises as the blade approaches and falls after it has passed by. Thus since fluids are elastic, pressure waves are constantly being created and propagated. Pressure waves which can be detected by the ear are called sound. If the combination of pressures and frequencies is pleasing, the sounds may be considered musical. If they are unpleasant, or just unwanted, they are called noise.

The intensity of the sound produced by a moving blade depends on its speed and its fundamental frequency depends on the product of its angular speed and the number of blades on the rotor. Sound patterns are complicated by the interaction of one blade with the sound waves produced by the earlier one and also by mechanical vibrations of the blade itself. The aerodynamic and the elastic properties of a blade can interact to produce an unpleasant and potentially dangerous type of oscillation, known as "flutter" in aircraft work. A high-speed windmill almost by definition can hardly be quiet and it may make a noise which is distressing. Low speed aerofoils need make little noise. Sail mills, being low speed, can be relatively very quiet until "slapping" begins on overspeed.

3 Terminology and characteristic features

Sails, sweeps and blades

The device which can bring about a transfer from the kinetic energy carried by the wind to power available on a rotating shaft may be some form of airscrew. The term airscrew suggests that the air is a solid medium through which the screw can advance when rotated. The term is accepted though the concept is far from the reality of a fluid/solid interaction. Among the devices commonly referred to as airscrews are windwheels, propellers and fans. The term sail is also a comprehensive one. It is the name which can be given to one of the elements of an airscrew as well as to the sails of a boat. It is also sometimes applied to the wings of aircraft and of birds. For consistency the author proposes that the term should be applied to a sail similar to that of a boat, that is a compliant textile fabric flying from a spar and restrained from one corner by a rope (at sea this rope is always called a sheet). The elements of the airscrew used on the traditional type of windmill, which are made of wooden boards or alternatively of canvas spread tightly over a lattice grid may be called sweeps. For a windmill which resembles an aeroplane propeller, or for a fan, the elements will be referred to as blades.

The windwheel

The whole rotating assembly of sweep, blades or sails which make up an airscrew may be called the windwheel. This is a more comprehensive usage than that of the old millwrights. The object of the windwheel is to slow down the wind and to alter its rotation (or swirl) in order to rotate a shaft. This can be done in many ways. A two-bladed, high-speed machine (or even a single-blade one), may be as effective in this respect as a multi-bladed low-speed machine, but it does not follow that the two are interchangeable. Always it must be remembered that it is the whole swept area of the windwheel disk that is the area which gives a measure of the available power and not the area of the blade elements themselves.

Pitch

The pitch of an ordinary screw is the amount which it advances through its nut in one complete revolution. This is necessarily the same at all radii. In an airscrew where the "nut" is not solid, the pitch is not necessarily the same at all radii and it may be defined arbitrarily at a particular radius. The pitch angle to be referred to in this book is the angle between the chord of an aerofoil airscrew and its plane of rotation. Unless it is specified otherwise it will refer to the angle at the tip of the blade. One pitch angle, common to all radii, is common in windmill work, but not so common in other applications.

"Solidity"

When a number of blades follow one another the ratio of the blade width to the distance from the centre of one blade to that of the next is called the "solidity". In a windmill context the measurements are made along a circular path and the radius has to be specified. Extreme values met with at the periphery of the wheel vary from something near to unity in an American type of mill to about 0.03 for a two-bladed aerofoil.

Support

There are three basic methods by which a wind power plant can be supported. These respectively depend on the concepts of the post, the mast or the tower. The post depends for its viability on being set into the ground and having sufficient rigidity to withstand the bending actions imposed upon it. Of all methods the post is the most economical in its use of ground space. The mast depends for its stability on the use of guy lines. This is potentially the most economical of material. The tower depends upon its weight which must always act within its base. In the case of a light lattice tower the significant weight is in the concrete, or other heavy footings to which each corner must be secured. The tower type of structure either in solid masonry or in wood or steel lattice has hitherto been much the most common.

The windshaft

The shaft which carries the windwheel will be called the windshaft. This is consistent with traditional usage. If the windshaft and its bearings have to be carried on a sub-frame then this may well be called the headstock. Headstock is a term used for the head or extremity of a machine consisting mainly of one important rotating member. The headstock of a lathe or of a pit-head winding gear can be cited in support of this usage.

Bowsprit

An extension of the windshaft forward of the windwheel may (using a marine analogy) be called a bowsprit. The bowsprit provides a forward fixing point for stays, generally to the blade periphery, which can strengthen the windwheel structure. These stays are rare, though not unknown, in traditional western mills but they are an almost universal feature of the Mediterranean sail mills. They have also been incorporated in the more successful of large experimental mills which have been built to augment public electricity supplies. Since such stays necessarily offer parasitic resistance, their existence can be regarded as a concession made by aerodynamics to structures.

Windseeking

Windseeking, that is, the means for keeping the windwheel facing into wind, and the additional problem of transmitting mechanical power from an elevated shaft down to ground level, was a basic problem faced by early millwrights. This was avoided in the southern Aegean by attempting only to use the prevailing winds, hence the monokairos ("single-weather") windmill of Crete. Fixed direction mills of a very different type are also reported from Nebraska, but in north-western Europe they are recorded only as folk memories. For a long time the problem was dodged rather than solved by turning the whole mill, including the machinery in it, to face the wind. The post mill thus evolved. In this concept the whole structure was balanced on the top of a post where it was located by a pintle. The post must of necessity go through much of the mill structure in order that the support could be above the centre of gravity. Furthermore, since the post was fashioned from a natural tree, there was a limitation as to size. In a primitive application the post could be let into the ground (the Tjasker mills of Friesland are supported in this way) but the disadvantage of the point of maximum stress being near to the surface of the ground must soon have become apparent. It is at the surface where air and moisture are present, that organic agents of decay are most active. Surviving post mills are commonly supported by a timber trestle which in turn is elevated on masonry piers, the whole support being enclosed in a round house. The importance of the central post was reduced in the Paltrok type of mill which was applied mainly to sawmills in the Netherlands. Here the mill was built on a circular turntable supported by rollers which ran on a circular track. The post remained as a locating device. Hollow post mills also occurred, in which a vertical shaft transmits power down the centre of a hollow post to machinery below the trestle. These survive in the Wip mills of the Netherlands.

The principal development was the use of the turntable instead of the pivot. This opened the way to the much larger tower mills where the turntable is elevated nearly to the top of the structure and carries only the headstock and the cap. The post mill has never been reported from the Aegean. Evolution apparently passed directly from the monokairos to the tower mill.

The floating mill

A most elegant solution to the problem, that of floating the whole mill on a barge or pontoon, has been recorded as an isolated example in the Netherlands. The possibility of a drainage mill being mounted on a circular concrete float, appears to have, at any rate, superficial attractions. Stability would be easy to achieve along with negligible friction of movement and an absence of foundation works. It could yield slightly to heavy gusts and thus relieve the stressing. Perhaps it may be considered if our wind power is taken seriously again.

The rotation of the mill to face the wind can be done manually or automatically. Manual adjustment was common among traditional windmills, but in Britain particularly, the fantail was largely used. This method, attributed to Meikle in 1750, provided a small windwheel set at right angles to the mainwheel, and on the downstream side of the mill cap, and was so geared as to turn the main mill around, which it did until it wound itself out of the wind. At this point the main axis would be receiving the maximum effect. Nowadays this might be called an error-actuated servo-mechanism but the idea far predates this kind of terminology. On a moderate to small scale a tail vane, or steering rudder, may be used, or the wheel itself may run downstream of the mast. Downstream mounting, it may be noticed, precludes certain types of control gear which can, in a storm, turn the mill out of the wind. A desirable property for a tail vane system is stability. If the direction of the windshaft is in a state of oscillation output suffers and unduly high gyroscopic forces may be introduced. The tail vane works in the wind "shadow" of the wheel where the velocity can be relatively low and disturbed. The use of a vertical symmetrical aerofoil as the tail vane surface offset well above the axis of the shaft can be considered.

Transmission of electric power to the ground

The mounting of a dynamo on the headstock avoids the problem of transmitting mechanical power down the centre of the structure but brings in the apparently more simple transmission of electrical power. Two methods suggest themselves. One is the use of flexible cables long enough to allow the headstock a number of rotations before they are wound up and damaged. Sites may exist where the wind changes in direction as frequently in one sense as it does in the other. If this is so this method, with some supervision, might be viable. On the other hand, sites certainly exist where such a method would be quite impossible. In such a case sliprings and brushes must be used. These must be well engineered and protected as they work in an exposed and corrosive environment. A bad contact can break the circuit abruptly and as quickly throw the load on again. Such shock can destroy some of the working parts.

Inclination of the axis

The small inclination to the horizontal such as is common in traditional machines follows from attempts to meet both structural and aerodynamic problems. To increase its strength and stability the windmill tower is generally built with a batter, or inward slope, that is, the tower has the form of a cone tapering upwards. It is necessary for the sails to swing clear of the walls and for the centre of gravity of the windwheel-shaft and brake wheel assembly (which can weigh upwards of ten tons) to lie between the bearings. Both these objects are most easily achieved if the shaft is inclined. The aerodynamic reason is illustrated in Fig. 3.1, which shows the stream lines of a flow approaching a cylinder. (These are traced from a wind tunnel photograph.) The widening of the gaps between the streamlines shows reduction in velocity, and from the cube law there is a much bigger reduction in kinetic energy.

In the plane A-A which is at one tower diameter upstream of the tower the flow is hardly modified. But at plane B-B which is one half of one diameter upstream the stream lines are being forced apart. A windmill sail, swinging through this region nearest to the tower, would move suddenly from full power into an almost dead region and equally suddenly back to full power. The consequent shock loading is severe. This normally results in characteristic thumping sounds. Inclining the shaft is one way of increasing the sail-tower clearance without lengthening the windshaft and moving its centre of gravity away from its safe place between the bearings. Another way is

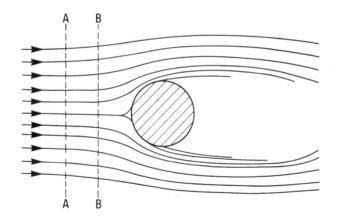

Fig. 3.1   Flow around a cylinder

to build the windwheel in such a way that its arms sweep out a forward cone instead of a disk, on rotation.

The conical or battered tower is almost absent in the corn mills of the Aegean and in Crete the windshaft is horizontal. The windshaft, however, is inclined in the Cyclades, and both windshaft inclination and batter are marked in the Algarve. However an illustration in the National Geographic Magazine, 1930, shows a cylindrical wood tower in Northern Spain. Rather unexpectedly in England, the mills of Cornwall and the West Country often had cylindrical towers, and one on Scilly had fabric sails, following the Portuguese rather than the Aegean rigging plan. This local use of a Mediterranean culture suggests a seaman's influence.[1]

The Tjasker mill

A specialised mill with a single shaft inclined at about 30° to the horizontal appears in the Friesland province of the Netherlands (Fig. 3.2). This is the Tjasker mill which represents an extreme example of simplification. The water raising device, known as the archimedean screw, necessarily has an axis inclined at about 30° to the horizontal. In the Tjasker, the screw shaft is extended until it is sufficiently high above the ground for a windwheel to be fitted. The Tjasker stands on a post in the centre of a little island surrounded by a low level drainage ditch. The water falls into an annular trough around the pole and is then conducted over a little bridge across the low level ditch to a higher level of drain.

Influence of duty on design

The type and appearance of a windmill can be much influenced by the intended duty since this prescribes the most suitable "torque-speed" characteristic.

A windmill which has to drive a reciprocating pump which in turn draws water from a deep well, needs to have a large starting torque, for the whole long column of water in the rising main must be set in motion. The torque can be allowed to fall off a little when once motion has begun. On a machine whose axis faces the wind useful effects come only from the lift mechanism, that is by bending the air flow. If the blades are formed to secure maximum bending (that is maximum starting torque) when they are at rest, they will assuredly be wrongly shaped for the best effect when they are in motion. Starting well and running well are quite different things. A high-speed mill which has

Fig. 3.2 The Tjasker mill

to generate electricity, being directly coupled to a dynamo, needs but a small starting torque for the dynamo does not accept its electrical load until it is up to speed. Here the requirement is for a machine which gives of its best when the speed is high. The torque should have risen to a maximum and just begun to fall as the working point is reached.

These duties represent extremes which are met by the American (or the Aegean) type on the one hand and by the high-speed aerofoil on the other. Alternatively, we could emphasise the difference of the objectives. The first aims to count up the greatest number of turns spread equally throughout the year, characteristically to keep a water tank topped up. It aims to use every little puff of wind that blows, but it can ignore most of the stronger winds when the tank will soon be full. The second aims to give a steady output of power when the wind is brisk and the mill speed is high.

The object of the windwheel is to change the velocity of the approaching wind in such a way that work is done on its blades. The approach velocity should be parallel to the axis. The leaving velocity will be lower and have a swirling or corkscrew-like motion. It seems fairly obvious that a fully bladed disk, such as an American wheel, can do this, but not nearly so obvious that a very narrow blade, cunningly shaped and moving at a high speed, can do the same as well or even better. Yet such is the case. The influence of a moving blade is transmitted both before and behind at the speed of sound itself, and the higher the speed of the blade the stronger is the distant influence. The author has an example of a high-speed two-bladed windmill which at running speeds almost completely stops the approaching wind. The area of the blades is only about five per cent of that of the circle they sweep through, yet measured speeds before and after passing through the disk have been observed at a ratio of 15 to 1.

High-speed, free-running windwheels have been used as small "windscreens" on the bridges of naval vessels in circumstances where the hazards of the moving blades were less than those of alternatives.

The number of blades

The obvious and important difference required to meet the contrasting duties is in the number of blades on the wheel. Modern versions of the American windmill might have twenty-four. Earlier ones which were built in a frontier situation where slats of wood were the most readily available blade material sometimes had as many as one hundred and sixty. When once the machine is running too many blades may interfere with each other's airflow so much that they limit the speed which can be attained and the power which can be developed. This is not always a disadvantage. Sometimes this aerodynamic imperfection is used as a protection against overspeed. The simple removal of many of the blades on a multi-bladed machine would probably increase the speed and it might well increase the power available but it would spoil the light weather performance and it might put the machine at mechanical risk.

Theoretical work by Professor J. H. Preston has shown that in an ideal case, where the blades do not interfere with each other's airflow and the blades are constantly adjusted for optimum mill output, then the speed of a windmill varies inversely as the square root of the number of blades. Thus a four bladed machine should run at about seventy-one per cent of the speed of a two-bladed, a six-bladed at fifty-eight per cent and an eight-bladed at fifty per cent.

The shape of the tips of the blades

Aerofoils discussed have been tacitly assumed to be very long in proportion to their widths. The flow is then said to be two-dimensional since only two component velocities at a point fixed by the radius and the axial position, and two co-ordinate dimensions, are needed to define it. Nothing changes along the length of the blades so that the third dimension is not effective. A real wing or blade is different in that it has an end which inevitably will influence the flow. An aerofoil builds up a high pressure on the under surface and a low pressure or suction on the upper. There is thus a tendency for the high pressure underneath to flow around the ends to the low pressure region above. This flow represents loss in energy and loss in lift. In aircraft such as gliders there is not much energy to lose and so the end flow is made relatively less important by using a very long and narrow wing. The same kind of shape appears in the high-speed windmill where the width of the blade is commonly about one tenth of its length. On other windmills the shape of the blade is influenced by the relative importance of easy starting and efficient running. Medium-speed machines have a width which is about one quarter of their length and very low speed machines may have so many blades as apparently to fill up the disk. Quite often, to ensure easy starting, windmill blades may increase in width outwards so that the width of the blade at the tip is greater than at the root. This makes starting more definite but it is at the expense of greater losses at the working speed. An exceptional case of windmill blades tapering to the tips (as aeroplane propellers often do) is recorded of the windwheel of the electric generator of an oceanographic-data buoy. The blades were made narrow and tapering so as to reduce the force of the seas which broke over them from time to time.

The centre of the windwheel

The middle part of the wheel, bounded by the half-radius circumference, only contains one quarter of its area. It can at very best, only contribute one quarter to the power. In a practical plant the middle of the wheel contains the hub, perhaps a gearbox and a brake, as well as the bearings and their supporting structure. Thus this centre part cannot be used and the contribution of the centre half must be less than one quarter of the total. It is also a slow-running region where blade angles are changing rapidly with radius and so is more difficult to design and construct. Whilst it is unlikely to contribute much power it would certainly contribute its full quota of overturning forces which the structure must withstand. In practice it is only in small high-speed machines that use is made of part of this area. Indeed the same arguments could be extended, without much power loss, to more than the area within the half-radius. Other rotary machines suffer in the same way and it is not unusual for axial-flow fans, turbines and pumps to use relatively short blades projecting from a large streamlined hub.

The special problem of the one- or two-blade mill

A two-bladed machine which can swing to face the wind, either under the control of a tail vane or by being mounted downstream of the mast, can be subject to a very special problem. Quite violent torsional oscillations can arise about a vertical axis. These reduce output by various shock losses and can be structurally damaging. They arise as an inevitable by-product of the change in moment of inertia about the vertical axis as the sail arms change from the horizontal to the vertical position. In order that angular momentum may be conserved the speed of swing must increase while the blades move from the horizontal to the vertical position and subsequently diminish as the sail arms become horizontal again. The severity depends on the relative masses of the swinging structure and of the windmill and on the angular speeds of both the motions. The problem is obvious in a windseeking mill but it may be latent even when the machine is clamped to face one direction.

Most engineering structures possess some degree of elasticity, hence a torsional mode of oscillation must exist. If the natural frequency of the tower were accidentally tuned to the rotational speed of the wheel, damaging oscillations could arise. Accidental "tunings" of engineering structures are not unknown; witness the Tacoma Narrows Bridge. The problem does not arise with a three- or a four-bladed machine.

Modern gearboxes

In recent years manufacturing technologies have advanced apace.

Modern mechanical production and the development of shaft seals have produced oil-bath gearboxes of incredible excellence and cheapness. Thus windmills capable of the extreme duties of water pumping and of electrical generation need not differ nearly so much. A machine with, say, six coarsely-pitched aerofoil blades and a reduction gearbox can cope with pumping, while one with three blades more finely pitched and a speed-increasing gearbox can cope with electrical generation. As well as oil-bath gearboxes endless flexible belts and their appropriate pulleys can be used to give the desired speed ratio in the final stage of a dynamo drive.

Classification

Classification may be regarded as the beginning of knowledge for it can lead to the discovery of order in the bewildering complexity of life. By noting the basic similarities in different objects the many can so much merge together as to become one, and the mind, when it can concentrate on one thing at a time, can study more effectively.

In the history of windpower, the number of types of machine which have been built is large and the number which have been proposed is larger still. Invention, study, travel and research constantly extends the list. Prospective builders of small windmills are often inventive and believe that they can open up hitherto unexplored possibilities in windpower. If a proposal is broken down into its elements by a system of aerodynamic classification which emphasises principles of operation (however much they may be hidden by mechanical detail), at the very least some measure of the possibilities and of the limitations can be determined.

Classification systems based on external appearances (such as the degree of taper in a windmill tower) have been applied to traditional mills. These have enabled early and often incomplete texts and illustrations to be correlated and they have thus cast light on the migration of ideas and cultures. This is of interest to the economic and the architectural historian. The engineer, by definition, is concerned routinely with power, strength and stability but his overwhelming consideration should be the "energy cost" of the structure. The successful machine is only that which repays in work the energy which went into its construction. This test is facilitated by a system of mechanical classification.

It is so very easy to suggest such things as variable pitch, aerofoil blades, air-brakes, slots, flaps and the like. The technology is available, but at a price. The justification for the use of windpower in a society such as ours is to save fuel energy and thus to make the world a pleasanter place to live in, also to keep alive skills and attitudes which may yet be of value. Normally it will be a mistake to erect a windmill which needs more energy in its construction than it can ever return during its working life, but even here there may be exceptions. An automatic, windpowered lighthouse on a distant rock can be credited with the energy saved by less frequent visits of the supply ship, as well as the power it produces, and thus justify a high initial cost.

Lift or drag?

The first and most fundamental decision in classification is whether it is the lift or the drag force which is to do work on the blades. To recapitulate: the lift force is that experienced by the blade in a direction perpendicular to the approaching stream. Thus before a lift or drag decision can be made some thought must be given to the effective flow direction of the approaching wind. Some windmills, for structural reasons, have a sloping windshaft even though they work in a horizontal wind. It is only the axial component (Fig. 3.3) of this

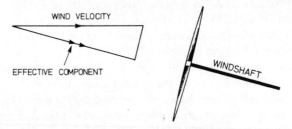

Fig. 3.3   Flow components of wind velocity

wind velocity which can produce the required lift force on the blades. Lift-dominated machines can also be made with an effective flow direction which is parallel to a diameter and thus perpendicular to the axis.

Ratio of blade-tip speed to the wind speed

When an aerofoil blade is used a range of efficient blade speeds is possible, depending on the form, number and pitch angle of the blades, thus another characteristic is introduced which is a measure of the speed. Speed is measured, not so much in absolute value, but as a ratio of the speed of a part of the machine to that of the approaching wind. Thus a high-speed machine may have a maximum tip speed up to ten times greater than the wind speed (quite exceptionally up to fifteen times). Such machines, in small sizes, might be directly coupled to an electrical generator without the interposition of gears, thus avoiding mechanical losses. Medium-speed machines have a tip-speed ratio of between two and five times the wind speed. The traditional mills of Western Europe were near the lower end of this range. Any efficient tip-speed greater than unity is necessarily influenced by lift but machines which have a ratio of unity or less may be dominated either by lift or by drag. Examples of low-speed lift machines are the American wind pump and the Aegean irrigation sail mill, both of which have a ratio of less than unity.

Effective flow direction

Three cases can be specified (Fig. 3.4). (A) is the axial flow which is common to all forms of airscrew, (B) is diametral flow which occurs in such devices as the cup anemometer, and (C) when part of the wheel is shielded and can be described as tangential flow.

Classification up to this point is represented by the diagram of Fig. 3.5. Group 1 includes the great majority of the real-power-producing windmills of the world. Group 2 includes some experimental devices such as the "venetian blind" windmill, and Group 3 includes the Darius design and its derivatives.

Drag machines appear in Groups 4, 5, and 6. Drag-generated work arises when the blade moves in the same direction as the wind but at a

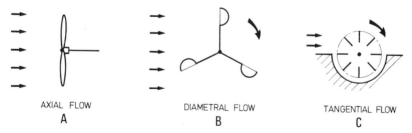

AXIAL FLOW   DIAMETRAL FLOW   TANGENTIAL FLOW
A   B   C

Fig. 3.4   Axial, diametral and tangential flow mills

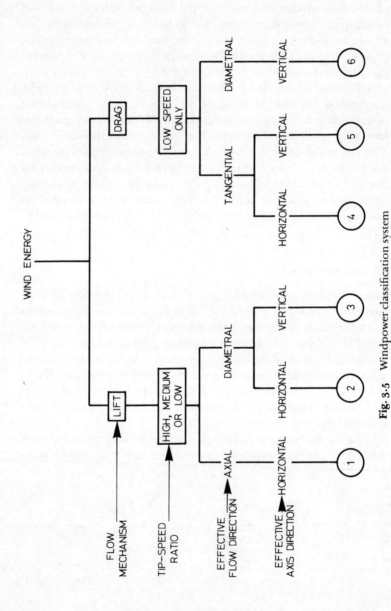

Fig. 3·5 Windpower classification system

lower speed. Since the blade must somehow get back to its starting point it must, for half the time, be moving against the direction of the wind. Thus the net force available is the difference between that experienced by the blade on its downstream and its upstream journey. The drag difference may be brought about by difference of shape between the front and back of the blade, as in the cup anemometer (Fig. 3.4) or by shielding one half of the wheel from the wind. These two types of machine may be called the differential drag and the shielded paddle respectively. The Group 4, horizontal axis, shielded paddles were formerly reported from Nebraska (i.e. the Nebraskan Go-Devil) and vertical axis ones (5) from Siestan. These latter are believed to be the oldest windmills on earth. One vertical axis differential drag machine (Group 6) is the above mentioned cup anemometer, a device established since 1846. The Savonius rotor which also comes in this Group 6 is used as a current-meter deep in the ocean. A minute version of the shielded paddle has been developed as a yacht log (or speedometer). The shielding arises here from its being recessed into the hull with only the blade tips emerging.

It is interesting to note that lift machines are, and always have been, dominant in real power production; drag machines (or very low speed lift machines) are dominant in instrumentation.

Classification here, in respect of windseeking, can range from "not sought for", which is the situation where the useful wind comes always from one direction, to "not necessary" when the machine (essentially on a vertical axis), can accept wind from any direction without adjustment. In between are manual devices where the whole or part of the mill must be winched into the wind by hand, and mechanical devices where the mill automatically follows the wind. These may be subdivided into fan tail, tail vane, and downstream rotor.

Inclination of the axis

This can be specified relative to the horizontal. The principal cases are horizontal, which can be regarded as normal for axial flow machines, and include small angles of inclination which are of both structural and aerodynamic significance. Among vertical devices are many which work equally well whatever the wind direction may be. These items are dealt with in Chapter 6.

4   The aerofoil

Streamlines are mathematically designed concepts invented to aid the analysis of fluid flow. When the flow is steady (that is not varying with time) streamlines coincide with filament lines in the fluid. Experimental techniques exist whereby these latter lines can sometimes be made visible. Fig. 2.4 (page 24) is traced from a laboratory photograph in which some of the filament lines (which in this case coincide with streamlines) around an aerofoil have been made visible by smoke. It is fundamental that flow cannot cross a streamline and so the passages between those made visible must adjust in width to suit changes in velocity. The streamlines over the aerofoil crowd together, hence the velocity, and with it the kinetic energy, must have increased. The extra kinetic energy can only appear at the expense of the pressure. Thus the closing up of the streamlines means that there is a fall in pressure over the aerofoil. Similarly their divergence below means a rise in pressure.

The basic property: large lift, small drag

The outstanding property of the aerofoil is the ability to produce a large lift whilst only incurring a small drag. To do this there must be little or no separation from the upper surface so that the benefit of suction can be realised. The negative pressure (or suction) on the top is more important than the positive pressure underneath. Secondly the turbulent wake shed from the trailing edge must be a minimum. That is, the parted streams above and below the aerofoil must come smoothly together as they leave the trailing edge (Fig. 2.4). Whilst the aerofoil must really be considered as a whole, we may (to aid understanding) say that it is the exact form of the upper surface which largely controls separation and it is the nose radius which gives some tolerance for changes in angle of attack. The fine trailing edge angle helps to diminish the wake.

Some of the benefits of suction lift can be obtained from a piece of sheet material, metal or plastic, curved to something like the upper surface of the aerofoil. Whilst effective in this one respect, the sharp leading edge is not tolerant of the change of incidence angle which occurs with changes of speed. Such blades are often used in pumping mills where cheap and robust construction, along with easy starting, are the dominant considerations. Indeed their imperfections as aerofoils (which increase with speed) may be aids to control.

The geometry of the aerofoil

The undoubtedly beautiful curves of the aerofoil follow from its functional perfection. Aerofoil shapes are calculated with a view to controlling the rate of pressure change in the boundary layer, for it is this which influences the stability of the flow. The calculations and the theories on which they are based are abstruse. Even so aerofoils can be classified by a relatively small number of geometrical parameters. Characteristic features of a hypothetical symmetrical aerofoil are illustrated in Fig. 4.1. These are (1) the nose radius, (2) the position and the magnitude of point of maximum thickness, and (3) the angle of the trailing edge. The first three characteristics are expressed in terms of the chord length. The upper aerofoil illustrated is symmetrical. It would not produce any lift unless the oncoming flow was inclined to the chord line. This inclination is called the angle of attack. Symmetrical aerofoils are used for control surfaces such as rudders and keels and they are essential features of lift-dominated vertical-axis windmills such as the Darius and its derivatives. Their special properties have also led to their being employed on certain large axial-flow windmills. In all the applications the angle of attack is that produced by the combination of the wind and the blade velocities. If an aerofoil is to develop lift at zero angle of attack, then it can be modified as in the lower diagram of Fig. 4.1. The point C is lifted above the chord line AB by an amount called the camber and a new line ACB is drawn. This is called the camber line. The aerofoil is the set-out equally above and below the camber line. This produces an aerofoil different in appearance but related geometrically to the symmetrical foil from which it was derived.

Whilst the general characteristics of the foil might be specified, the exact shape of the enclosing surface and the form of the camber line

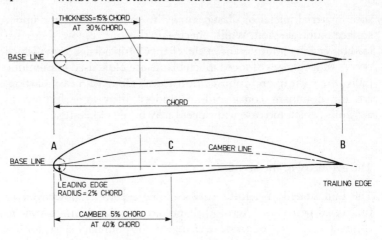

Fig. 4.1   Geometry of symmetrical and cambered aerofoil

depend on the mathematical philosophy of the designer. The illustrations, whilst meant to be broadly representative of low-speed aerofoils suitable for windpower, are not meant to be exact representations of any particular design. Thickness adequate to accommodate a substantial spar can be an important consideration in the choice of a windmill aerofoil.

The properties of the aerofoil

An aerofoil design is put to proof in a wind tunnel and if it shows sufficient promise its properties may ultimately be published, among hundreds of others, in appropriate books of reference. Wind-tunnel models are made with a high degree of accuracy and perfection of surface finish. Less accuracy has often to be accepted in the field but it must be realised that only a small imperfection in an aerofoil can trigger off separation and result in a performance different from that expected.

Properties of primary interest to the windmill builder are the lift and the drag, and sometimes the turning moment, which the flow exerts on the aerofoil. A systematic series of experiments would normally be made over a range of values of the angle of attack, the wind tunnel speed being held constant. Fortunately, over the practical range, mathematical methods exist whereby, when the properties at one speed have been observed, those for many others can be calculated. Curves of lift and drag, plotted against the angle of attack for a low-speed aerofoil, are shown in Fig. 4.2, along with another curve which is the ratio of lift/drag. The lift and the drag are shown in

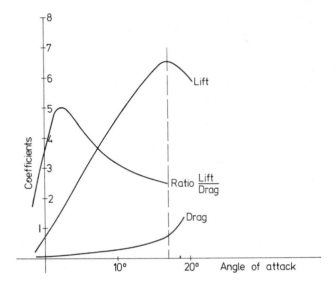

Fig. 4.2   Properties of an aerofoil

coefficient form. At a given wind speed it can be seen that a non-symmetrical aerofoil has to be held a few degrees nose down, or have a negative angle of attack, to show zero lift. As the angle of attack is increased, the lift increases steadily at first, then at a progressively diminishing rate until it reaches a maximum and begins to fall. As the lift approaches its maximum the drag begins to increase at a much greater rate. The fall of lift and increase of drag indicate the breakdown of ordered flow over the aerofoil which is manifested in the stall.

The stall occurs when the angle of attack involves a curvature of flow which is more than can be sustained. (Fig. 2.5, page 25.) In flight the stall can be catastrophic, particularly at low altitudes when the pilot has little room for manoeuvre. In a windmill application it merely makes the windmill hesitant in starting.

Use of coefficients of lift and drag

The quantity of aerofoil data which could be recorded is large, since for each form there is an indefinite number of combinations of speed, density, size and angle of attack. To make such data compact and easily available special types of plotting are devised. These depend on the assumption that for any one form of aerofoil at one angle of attack the lift and the drag both depend significantly on (a) the square of the velocity V (b) the density of the air ρ, and (c) the area A of the aerofoil. These assumptions are justified by experience over the range of conditions likely to be encountered. Another tacit assumption regarding the area A is that the aerofoil has a reasonable length in comparison with its width.

The lift and drag are then given by the equations Lift $= C_L (\tfrac{1}{2}\rho V^2)A$ and Drag $= C_D(\tfrac{1}{2}\rho V^2)A$, where C_L and C_D are known as the coefficients of lift and drag respectively. C_L and C_D are now the basic values to be plotted on the graphs against the angle of attack. To find the lift or the drag at any particular angle of attack the appropriate coefficient can be read off the graph. This, multiplied by $\tfrac{1}{2}\rho V^2A$ gives the value of the force. The figure $\tfrac{1}{2}$ is included because $\tfrac{1}{2}\rho V^2$ has an easily interpreted physical significance. It is known as the stagnation pressure.

"Runaway"

The lift/drag curve shows a peak at the aerofoil's optimum operating point but it also shows that the aerofoil can operate for an appreciable range on either side of the optimum point. Indeed the aerofoil windmill may work sufficiently well over a range of angles of attack up to about 15°. This can lead to problems of overspeed on loss of load. When the aerofoil is finely pitched for high tip-speed ratio then, even when the speed is indefinitely high, there may still be a component of lift force trying to make the mill go faster still. However, in a real machine friction is always present and there will be a natural limit to speed when rising friction catches up with the driving force. It is only in small sizes that this balance may be attained with safety. The danger of runaway is an unavoidable penalty associated with a finely pitched high-speed windmill. It is generally met by fitting an automatic brake. Conversely, if an airscrew type of device is to be used as an anemometer, it is desirable that the pitch should be a coarse one and the section should be more like a flat, thin plate than an aerofoil.

The limits of speed

The high-speed aerofoil windmill coupled to a generator came fifty years ago as a by-product of aeronautical progress. These machines are slow to start but when movement begins the speed increases, slowly at first, then more and more rapidly as the aerofoil comes out of stall. These machines are a limited and drastic solution to the generator problem. A hypothetical 2 m (6½ ft) diameter machine which turned at a rate of 20 times a second would have a tip speed of 125.6 m/sec (281 mph) approaching one third of that of sound. At this speed, Mach No. or compressibility effects are becoming measurable. At higher speeds they can become dominant and they can be manifested by vibration, noise and loss of efficiency. The limit of speed is the tip speed of the blades, not the rotational speed of the wheel as a whole. Thus the direct-coupled machine is necessarily a small scale solution only. Fifty years ago some early manufacturers in the United States used (and may still use) contra-rotating aerofoil windmills on the same axis. One drove the armature in one direction and the other drove the casing carrying the field magnets in the other direction, thus doubling the effective generator speed.

The hazards of high speed must not be forgotten. The speeds involved are comparable to those of an aeroplane propeller. A blow from such a windmill could be lethal and there is always the latent risk of runaway. Mechanical risks involved can be illustrated by considering a hypothetical 2 m (6½ ft) diameter machine rotating at 20 r/sec (1200 rpm). If a blade weighed 4 kg (9 lb) then the centrifugal force produced would be 32×10^3 N (or about 3 tons force). The shedding of a blade due to fatigue could release forces which would certainly destroy the structure, quite apart from the damage the blade could do as a missile.

The symmetrical aerofoil

The section on aerofoil geometry was illustrated by the development of a symmetrical aerofoil which had obvious application to control surfaces. In a windmill content it has certain special applications particularly when it is used without any pitch angle. Even though such an aerofoil has no inclination to its plane of rotation it can nevertheless produce lift and drag when it is subject to a side wind. This is illustrated in Fig. 2.6. It can be seen that forward motion of the aerofoil is essential to produce a relative velocity in the appropriate direction. Thus this kind of windmill is not generally self-starting and is normally provided with auxiliary power to run it up. However, the author has found that such aerofoils do sometimes start themselves, due to the different flow patterns which form around the blunt nose and the sharp tail. It would be imprudent to leave unbraked a windmill based on one of these aerofoils.

Symmetrical aerofoils may be used for their intrinsic properties both structural and aerodynamic. This was done by the late Professor D. E. Elliott on his large machine in the Isle of Man.[2] Another advantage arises when a windmill is coupled to an A.C. main. The generator must run at synchronous or near synchronous speeds and the windmill must keep on running, even if it is drawing power from the mains during lulls in the wind. The symmetrical aerofoil with zero angle of attack being almost perfectly streamlined, is easier to drive in such a case than is one which. acting as an outsize fan, is creating a wind of its own.

The advantage of the aerofoil

The possible gain in power and speed which can result from the use of an aerofoil section is illustrated in Fig. 4.3. A 2 m (6½ ft) diameter, two-bladed machine was fitted alternatively with low speed aerofoil blades (made to a high degree of accuracy and finish but with no twist) and with blades cut from sheets of 10 mm (⅜ in) plywood. The

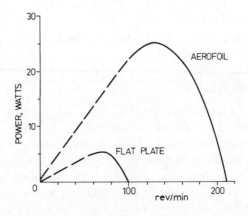

Fig. 4.3 Aerofoil and flat-plate windmill compared

plywood was square edged and treated with a coating of protective paint but with no special attempt to achieve a high degree of finish. The plan forms of the blades were identical and the pitch angle was $7\frac{1}{2}°$ in each case. The wind speed was 4.5 m/sec (10 mph). The difference was dramatic. The aerofoil rotated at twice the speed and developed five times the power of the flat blade.

The question arises, can the advantages of the aerofoil be sufficiently achieved with a freehand shape, produced by a craftsman, instead of one of the recognised and published mathematical forms? In a particular case the author would say that he has come across a windmill so produced whose performance could hardly be improved. On the other hand, why take a risk when proved shapes are available?

The twist (helix) of the windmill blade

Since the speeds, in rotation, of different points along an aerofoil vary with their radius, then if the "attack" is to remain constant along the blade, there must be a continuous twist from root to tip. For an aircraft propeller, where the weight penalty is dominant and no refinement is too great, this twist is normally used. A windmill does not have a severe weight penalty and thus aerodynamic imperfections, impossible to contemplate in an aircraft, may be acceptable aerodynamically and be desirable structurally. In practice, so variable are the conditions of operation and so generous are some aerofoil forms to change in angle of attack, that successful windmills can be built with no twist at all and with a fixed pitch. Indeed the absence of twist may impose little penalty on output, lend itself to cheaper and stronger construction and be a positive aid to regulation.

Acceptable angles of attack for lower aerofoil of Fig. 4.1 range from about 2° to about 15° in a windpower context. If this aerofoil is set at the optimum pitch angle at its tip and with a tip speed ratio of six, then at half-radius the angle of attack has increased by about 9° and is thus still in a favourable range. If the doctrine of the inadequacy of the centre of the wheel is accepted, there is little point in putting any twist into such a blade.

5   The sail and the sail mill

Mediterranean and European mills

The great increase of aerodynamical knowledge which came with the aeroplane has only been of limited benefit to windmill building, possibly even the contrary. When the glittering possibility of a great increase of power which aerodynamics can bring has taken precedence over the problems of structure and control, then disaster has surely followed. More realistic as a model for the windmill is the sailing ship, a concept which has been showing slow but certain evolutionary progress for at least four thousand years. Sail and sheet are uniquely adapted to abstracting wind energy in the stormy layer of air which covers our globe. With every Fastnet yacht race the evolutionary process inches on. In our own time we have seen the passing of the yard and gaff and the adoption of the high-aspect ratio sail, along with the completely triangulated structure of spar and stay. No structure more economical of material can exist. These facts have been known, if unwritten, in the islands of the Aegean and in parts of the Mediterranean for a long time but they are only just beginning to be appreciated in the scientific west.[3]

By sail windmill is meant the type whose windwheel consists of a fairly large number (four to twelve) radial arms from which fly fabric trisails in much the same manner as the mainsail flies from the mast of a yacht. These machines are common in the islands of the Aegean and in other parts of the Mediterranean seaboard. According to Wailes[4] they are never found very far from the coast. The bulk of those rendered familiar by travel brochures were built for corn grinding. Fifty years ago in response to local needs, a small variant developed in Eastern Crete. This was in effect a combination of the local type of sail mill and a simplification of the 'American' wind-pump incorporating a lattice tower (Fig. 1.6). They were used for irrigation and at the peak, perhaps fifteen years ago, many thousands were in use. They were restricted to the region east of Heraklion, the greatest concentrations being on the coastal plains of Mallia and the mountain plateau of Lasithi.[5] [6] It is interesting to speculate why they were not used in the western half of the island. The conditions around Rethymon appear equally favourable to those of Mallia but here animal driven chain-pumps were predominant before internal combustion engines and electrical power took over the local pumping.

A feature of these windmills is a forward extension of the windshaft which may well be called a bowsprit. From the outer end of each sail arm there is a wire stay to the tip of the bowsprit. These effectively take the bending moment away from the point where the sail arm is mortised into the windshaft. Further stays extend from tip to tip of the sail arms; these rather correspond to the triatic stays of a multi-mast sailing vessel. The function of these stays is to take charge of centrifugal forces and generally to stiffen the wheel structure. Each sail is sheeted to a point on the triatic stay, which in Crete at any rate is commonly made of chain, and the sheeting consists of engaging a hook attached to the corner of the sail in an appropriate link in the chain. When the wind is strong and reefing is required the sails can be wound around the poles so as to reduce their effective area. The next stage in a rising wind would be to remove some of them entirely. The author is not aware of automatic roller reefing ever having been applied to these mills but it could be done, given a little ingenuity and mechanism.

In Crete the inner ends of the sail arms are mortised into the windshaft on the traditional corn mills. The arms are thus of necessity spaced apart axially to avoid undue local weakening of the shaft by too many mortise holes close together. Even so the tips of the arms are constrained to lie in a single plane. Thus the sails are in effect, to use a nautical analogy, sheeted amidships. This is not the procedure normally carried out at sea when a high performance is needed. Some of the mills in Southern Portugal (and one which formerly existed in Isles of Scilly) use two planes of rotation. Four arms in the upstream set flew the sails which were sheeted to four other arms set further back toward the body of the mill.

Simplicity and self-regulation of the sail mill

The sail mill has many advantages as well as the great strength and low cost of the triangulated structure. The aerodynamic surfaces are

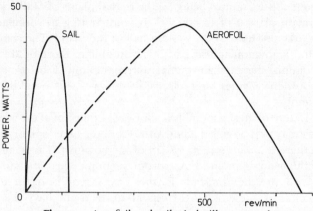

Fig. 5.1   Aerofoil and sail windmill compared

self-forming and it has a degree of self-regulation. A moderate increase in relative speed leads to a change of angle of attack which has the same effect as pointing a sailing boat too high in the wind. The sail is "taken aback" and so the driving force ceases. The author has put this to proof on a 4.2 m (14 ft) diameter mill exposed to a wind gusting up to more than 18.2 m/sec (40 mph) with the load removed from the shaft. The mill survived without harm.

A disadvantage of the sail mill for electrical power generation is its low speed, its blade-tip speed of about 0.8 of the wind speed being commonly experienced, though higher values are possible.

From experience in the field, the low speed efficiency can be high, but as with a yacht, where small differences of sheeting can have a dramatic effect on the outcome of a race, whilst it is very easy to set up such a mill for adequate results, there is scope for experiment to find the maximum performance.

The size of a Cretan pumping mill is generally about 4 m (13 ft) diameter. The corn mills are about twice this size, that is about one half the diameter of a large Dutch polder mill. The limitation of size of the Aegean mill is not inherent in the construction; yacht masts more than forty feet high are common. Since the stays on a windmill are at a more favourable angle than the shrouds of a yacht, there should be no technical difficulty in building a sail mill at least as large as a traditional mill of the west. Such a mill would begin to turn in the lightest winds and would be of the type that worked for the maximum hours per year. The best use of such a high-torque low-speed power is undoubtedly for water pumping. Ideally it would work through a centrifugal pump which by following a cube law would have an extra regulating tendency.

Successful applications and duties

Whatever the possibilities of the hypothetical large sail mill, a small one, of say 2 to 4 m (6½–13 ft) diameter is an extremely practical machine. It can drive simple workshop tools such as drill, lathe, or grindstone. A flywheel would help toward uniform speed in such a situation. Alternatively, one can drive a moderate power battery-charging generator but a considerable gearing up (of ratio up to about 60:1) is needed in such a case. An eight-sail machine 2 m (6½ ft) in diameter has driven an ex-wartime emergency hand generator very satisfactorily and a 4 m (13 ft) diameter machine has maintained an adequate low-voltage lighting system sufficient for two cottages all the year round. In each case the mill was excellently sited to meet the wind and there was adequate battery capacity. Such machines need some supervision and sail changing according to the wind. The danger of failure of the sails is due to flapping, rather than from excessive speed, it is interesting to note.

Sail and aerofoil compared

Extremes of properties are illustrated in Fig. 5.1 where the power-speed curves for two windmills of the same diameter and under the same windspeed are shown. These curves were taken by the author in the laboratory from two models 1.5 m (5 ft) diameter under a windspeed of 5 m/sec (11 mph). One machine had eight sails, which filled up 40 per cent of the disk area, these being rigged for optimum performance. The other mill had two thin aerofoil blades 0.075 m (3 in) wide. The sail windmill could develop 42 watts at 75 rpm while the high-speed mill developed 45 watts at 450 rpm. The starting torque of the aerofoil was so small that it sometimes needed assistance to get away. In comparison the starting torque of the sail mill was very large.

The rigging of a sail mill

A sail mill has many variables; the number of arms, the size and shape of the sails, the tightness of the sheets (the cords securing the free corners) and the points of their attachment. In practice it is so tractable a machine that it is easy to build one to give a tolerable performance. However, by controlled experiment the author has been able to obtain power outputs comparable with those given by the best aerofoils at his disposal, albeit at only a fraction of the rotational speed.

A guiding principle in the setting of the sails is that it is important for the spent air to be able freely to leave the windwheel without obstructing the flow to the following sail.

When maximum starting torque is desired (as it would be for irrigation in a region of light winds) then fairly loosely-sheeted sails flying from all the arms of the mill would probably give the best results. When maximum power is required in a moderate wind, as in an industrial or electrical-generating application, then it may be advantageous to reduce the number of sails in use, or to use the same number of smaller sails. When this is done the starting torque will be much reduced but, given careful experiment to find the optimum sheet length, the maximum power and the speed at which it is produced may both be significantly increased. So critical does the sheet length become that the author prefers to use light chain instead of cord for the sheets to avoid slipping or stretching.

As a first trial in setting the sheets the author would suggest a sheet of such a length as to allow each sail to stream downwind by an amount equal to one-fifth of the breadth of the sail. Longer and shorter sheet lengths can then be tried until the optimum point is reached. (Detailed experiments are reported in "The characteristics of a sail mill".[7])

6 The vertical axis mill

There are examples in history of inventors working on projects which appeared so impractical as to excite the derision of their contemporaries. In due course, however, advances in knowledge of materials, and manufacturing techniques, have removed obstacles to progress. Former folly then became accepted practice. Such may be the case of the vertical axis windmill. Until recently it seemed that inventive effort spread over hundreds of years had led only to the cup-anemometer, the rotating chimney-pot, and the Savonius rotor. Recently, appreciation of the properties of the symmetrical aerofoil has opened a way and large-scale experiments on real power producers are in progress.

Apart from these recent experiments the fascination which a vertical axis windmill can exert to newcomers to windpower studies cannot easily be explained. Certainly some types of vertical axis machine are able to accept the wind from any direction and thus to avoid one kind of complication. This advantage may be dearly bought in terms of output and it removes any chance of turning out of the wind when it is necessary to ride a storm. Over the years it has attracted a lot of effort which, when seen in retrospect, could have been much better directed. This tendency continues today. Indeed, so fascinating does the concept seem to be, that one could almost suspect that the attraction is something stored deeply in the mechanical section of the subconscious mind.

To put the vertical shaft windmill into perspective we may look at the earliest devices of quite different kinds in which rotation has been applied in the service of man. This will take us back to the ancient world.

Stone vases, such as those which can be seen in the archaeological museum at Heraklion were bored with a hollow drill fed with an abrasive powder and water. The drill was rotated with a bow and for ease of working and for control it would be worked vertically downward. The potter's wheel rotates in a horizontal plane. Ancient examples of these can also be seen in the museum and some which are not very different can be seen at work today. Most corn grinding devices, developed perhaps from the pestle and mortar have the same property. Indeed to an engineer of the ancient world rotary motion would be instinctively associated with the vertical axis. I am sorry not to be able to suggest that Daedalus built a helicopter to facilitate his abrupt departure from Crete!

When, perhaps some two thousand years ago animal power was first being used for the grinding of corn the vertical axis would still be the natural one to use, it being much easier to constrain an animal to a circular path on the ground than to put it in a treadmill.

The mill to which Apuleius was harnessed during his unfortunate transformation into an ass would have a vertical shaft directly coupled to the upper millstone. From this vertical shaft there would be two radial arms; between the outer ends of these arms the blind-folded donkey would be attached head and tail respectively. In such a way the unfortunate animal was constrained to walk along a circular path. (The construction of this part of the mill could hardly have differed much from some of the donkey-driven irrigation pumps which are used in South West Portugal today.) The earliest water mills were quite probably vertical axis machines (as are the largest hydro-turbines today). Very simple and effective vertical shaft water wheels can still be seen at work in Crete, in Portugal, and in Madeira.[8] One marvels that such silent and effective machines can be built from such a limited raw material. The vertical axis eliminates the need for a right-angled drive in the corn mill (but not in the chain-of-pots irrigation pump). What may be the earliest form of windmill reported from Siestan worked in this way; in effect it was a large paddle wheel on a vertical axis with one side shielded from the prevailing wind.

One can read of many attempts at building vertical shaft wind-mills written by historians concerned with the history of engineering. Here a word of warning is necessary. Some writers in the last century define a wheel by its plane of rotation. Thus the vertical axis machines are sometimes called horizontal windmills. Mathe-maticians, astronomers and modern engineers always define a rota-tion by the direction of its axis, and so much recondite theory is built into this that they will undoubtedly continue to do so. To avoid am-biguity the rather longer description of "rotation about a horizontal axis" or "rotation about a vertical axis" will be used. Sir William

Fairbairn, the Victorian engineer (associated among other things with the former tubular bridge over the Menai Straits) writing over one hundred years ago remarked, rather severely—

"Much has been done and a great deal of labour has been expended in endeavouring to improve the horizontal mill (i.e. the vertical axis) but without success. In fact, the horizontal windmill requires so large a surface exposed to the plane of motion, while moving in a material of the same density by which it is impelled, that it suffers great retardation, and from the principles of construction and the position of the vanes only one or two can be in action at the same time."

It is extremely easy to get a device which will, in a wind, rotate about a vertical axis. To get useful work out is quite a different matter. There is a record of a steam locomotive, which under the influence of a Pennine gale, went round and round on its turntable quite out of control. Ultimately it was slowed down and stopped by ashes being shovelled onto the turntable track.[9] It was probably acting as a shielded paddle wheel or it was picking up vortices arising from the dreadful winds which can speed down from Cross Fell. It is interesting to speculate whether a mistake in the design of a weathercock has ever led to it going round and round instead of pointing into the wind. Thus it is that variations of the vertical axis windmill which accept wind from any direction (and some which do not) continue to attract re-invention at the present time. In practice the advantages often prove more apparent than real in that energy developed is often rather small for the size and cost of the structure. Fairbairn's severe strictures of 1865 have not altogether lost their validity.

Recent developments

What may prove to be the practical breakthrough is the South-Rangi windmill which has recently been developed in Canada. This employs three symmetrical aerofoil blades which whirl, like vertical skipping ropes, about their axis. The blades are able to take up the natural form of the catenary (modified only by gravity) the consequent limitation of bending stresses leading to economy of material. They are necessarily high-speed machines and as such they need precise balance and refined bearings. In its simple form it might not be self-starting, since in this context the symmetrical aerofoil does not develop lift until it is in motion. This however is not a very significant problem.

From published work the output for a given swept area might well be less than that of a propeller-type machine since part of the rotation must be parasitic. On the other hand its output for a given weight of constructional material could be better. The outcome is awaited with interest.

The "Magnus effect"

The fact that a ball could be made to follow a curved path, other than its natural parabola, must have been known to skilled exponents of ball games for a very long time and the fact that similar effects arise in artillery was certainly known to military engineers. As far back as 1794 the Berlin Academy offered a prize for the solution of this problem. In particular, a shell fired from a rifle barrel could be altered in range by a side wind, not so much laterally in the direction of the wind, but perpendicular to it. A shell with a right-hand rotation, when viewed from behind the gun, goes beyond its target when the wind is from the left and falls short of it when the wind is from the right. The effect was studied intensively by H. G. Magnus who was a professor of physics in Berlin and, appropriately, an instructor in the Artillery Academy. Magnus reported in 1852 and subsequently the aerodynamic forces which can be produced by the spin of an object in flight have been described as the "Magnus effect". Many years later Lord Rayleigh wrote on the more complex case of "The irregular flight of a tennis ball".

A simple qualitative explanation of the effect can be based on a picture of the streamlines (Fig. 6.1). On the side where the cylinder

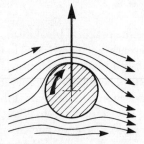

Fig. 6.1   The Magnus effect

surface and the wind are moving in the same direction, the flow is accelerated. Hence the streamlines are drawn closer together and the pressure falls. The opposite effects apply on the other side with the net result that the cylinder experiences a force perpendicular to the direction of the wind. This force can be of very considerable magnitude. If movement can be allowed in the direction of the force, then there is the possibility of doing work over and above that required to rotate the cylinder.

Flettner and the rotor

The first twenty years of this century saw the rapid decline of the sailing cargo ship. An attempt to arrest the decline was the auxiliary, a sailing ship fitted with a relatively small engine which was meant to get her in or out of harbour and to keep her moving through the doldrums and at other times when the wind was unfavourable. The method failed; the windage of the spars and rigging was too great for the engines used. A larger engine and associated fuel bunkers would have encroached so much on the cargo space that they could not be contemplated. In the nineteen-twenties a German engineer, Anton Flettner, was looking into the reciprocal problem. Instead of asking if a basically sailing ship could be helped on occasion by an engine, he asked if what was basically a powered ship, could be helped from time to time by the wind.

Anton Flettner, in his book,[10] describes himself as a teacher in the municipal schools of Frankfurt. He was also a prolific and energetic inventor, who perceived more deeply than most, the significance of the new fluid mechanics. During World War One he had been associated with problems of aircraft control and he had developed a servo-system by which a small auxiliary rudder could draw energy from the mainstream flow and use it to operate the main rudder. This was developed and used on all large German aeroplanes at that time. After the war he obtained permanent leave of absence from the teaching profession in order to further his inventions. For a while he went to live in the Netherlands where he found support for the marine applications of his ideas. Subsequently he returned to Germany. Whilst his principal activity was the development of his rudder for marine purposes, his abiding, and almost obsessional, interest was in wind power. As a youth he had sailed the windjammer route to Australia and he realised that wind power applied to a ship was to be measured in thousands of units, while any application on land was a matter of tens and units.

Back in Germany he received a contract to develop a sail based on a rigid symmetrical aerofoil controlled by his .sual auxiliary tail vane. During this work he commissioned Pran ltl's laboratory at Göttingen to carry out model tests. On a ro itine visit to this laboratory he came across the Magnus effect whicl was being studied for its intrinsic interest, without any particular application in mind. Shortly afterwards, Flettner tells, the view came tu him, almost with the certainty of a religious conviction, that the Magnus effect was the key to the use of windpower both at sea and on land. He was on a seaside holiday with his wife at the time but he returned to work immediately, first to convince himself further with a model test and then to persuade his sponsors to transfer their support from his metal aerofoil sails to the use of a Magnus effect on a spinning rotor. Instead of a ship surmounted by spars and billowing sails (with their great air resistance when under power) Flettner envisaged a ship mounting gigantic spinning cylinders driven electrically from a relatively small diesel generating plant.

The rotors certainly looked ungainly and unseamanlike but their area was only one tenth of the sails they replaced, and Flettner calculated that their weight and static wind resistance would be less than one quarter of that of the spars and rigging which they replaced.

Instead of setting, sheeting and reefing sails according to the wind, all that would be required was bridge control of the direction and speed of the rotors. Thus, Flettner argued, wind power could be used to the best advantage and without appreciable delay in meeting changing conditions.

A limitation which he does not mention (but which must undoubtedly have occurred to him) is that a rotor can only use the component of wind which is on the beam. Not only must the rotor ship tack when the wind is dead on the nose, but when the wind is from the stern as well.

Flettner's sponsors accepted the change and in due course basic studies and model tests were carried out under Prandtl at Göttingen. When these were completed came the time for full scale sea trials for which purpose the "Buchau" was chartered. This three-masted auxiliary schooner, of six hundred tons, was dismasted and fitted with two rotors, each 20 m (66 ft) high by 2 m (6½ ft) in diameter. These could be spun at speeds up to 200 rpm by electrical power derived from a 45 hp diesel engine. Her auxiliary propulsion engines remained. Her maiden commercial voyage as a rotor ship was from Danzig to Leith with a cargo of timber. She arrived on 5th August, 1925. The voyage had taken twelve days and the propulsion engines had been used during heavy weather in the Baltic. Presumably they were also used in the passage of the Kiel Canal. Otherwise we assume that she sailed under her rotors. Flettner reports (quoting Lord Byron), "I awoke one morning and found myself famous". Inevitably he suffered from undue publicity. Meanwhile "Buchau" had been leased to an exhibition contractor and used for (implied) less worthy purposes, such as amusement and dancing. In due course she was purchased outright by Flettner's group and renamed the "Baden-Baden". As such she crossed the Atlantic by the southern route (as most sailing ships still do). On this voyage she was only under rotor power alone for two hundred miles. At other times it was considered that on average the rotors were making a twenty per cent contribution to her propulsion.

Another ship, the "Barbara", of 2,800 tons, was specially built as a rotor-assisted ship. She had main engines aggregating one thousand horsepower, arranged so that varying proportions of engine and rotor power could be used with economy. She certainly went into commercial service but the author lacks exact knowledge as to how long she lasted as a rotor ship. There is an impression that it was for about four years.

The rotor aroused considerable contemporary interest. Enthusiast groups such as technical students built their own rotor boats and Flettner tried to initiate a rotor yachting movement. However, subsequent interest gently faded away. Perhaps the economy in propulsion derived from the rotor was offset by the cost and the loss of cargo space in the commercial field, and it only needs a little knowledge of the yachting world to realise how unacceptable such an unorthodox hybrid of wind and power would be.

As the "Buchau" was berthing in Scotland, Flettner was at work on the rotor windmill. Each of the windmill arms carried a spinning rotor, whose Magnus effect caused the main wheel to rotate. Test models and prototypes were built, culminating on one sweeping an 18 m (60 ft) diameter, supported by a tower 30 m (100 ft) high. The rotors were 5 m (16½ ft) long, tapering in diameter from 0.9 m (3 ft) to 0.7 m (2¼ ft) at the tips. This machine was later reported to have worked for four years, and larger machines were being planned. How long this machine lasted, how long and how well it worked, what is to be learnt and what avoided by future experimenters, the author does not know.

An authority on big windmills has told the author that the Flettner method must still not be dismissed. A great danger to the big windmill is runaway if it loses its load. All the power of the wind is in an instant turned to destructive acceleration. Automatic safety devices, coning, feathering or centrifugal brakes, can, if they work unevenly, throw things out of balance and so add to the destruction. The failure of the power supply to a Flettner rotor can only make it safe in that it cuts off the creation of power at its source. Flettner himself said that to concentrate attention on output at the expense of regulation, which is what the high-speed aerofoil windmill projects must do, was the way to disaster. He has not yet been proved wrong.

Savonius and the wing-rotor

Engineer Captain S.I. Savonius from Finland strongly believed in the potentialities of the rotor but saw little hope of their being fully realised whilst it depended on an external source of power. This he sought to rectify by making the rotor itself a multi-directional windmill. The wind was first to spin the rotor on its own axis. This would then produce a Magnus force which was to be the principal means of doing work when motion began.

Savonius' experimental method was ingenious. Having found one form of rotor which would spin in the wind, he then made another of slightly different form and set both in opposition on the same shaft. If there was much difference in performance one would overcome the other and drive it backwards. The poorer one was then rejected and the process repeated. Thus by successive steps, in which he varied the number and the shape of the blades, he arrived at his well known "wing rotor" (Fig. 6.2), a cylinder split longitudinally and the two sides given a relative displacement. By opening or closing the gap between the wings the properties of the rotor could be varied and it could be reversed in direction. It is not clear from Savonius' work whether or not he himself realised a limitation of this method. His experiments certainly led to the rotor with the best starting torque, but this is not necessarily the same as that with the best running performance.

Fig. 6.2   The Savonius wing-rotor

Savonius applied the Magnus force of his rotor to boat propulsion where he was able to exploit its reversibility in tacking. He also built a windmill with wing rotors on its arms. This he found could exert a large torque at a low speed. Then he began to use his wing rotor as a prime mover in its own right without regard to its Magnus effect. His enthusiasm for his work shines through his writings even now.[11] In one photograph he stands at attention, dressed in tropical kit, on the end of a pier, his experimental windmills all around him. Also clear in his writing is in the fact, rather severely commented on by a contemporary, "His experiments whilst extensive are mostly qualitative."

By 1931, as the Western world slid into depression, the excitement produced by the rotor and the Magnus effect had died away, but Savonius' rotor had established itself for a duty it was well able to do, driving car-roof ventilators, a situation where it generally had a high-velocity wind able to compensate for its rather low intrinsic power. He was also able to report that he had used it successfully for the abstraction of wave energy on a sea beach. More correctly perhaps he had abstracted energy from the flow resulting from a decaying wave system. Water surges up a beach as a wave arrives and the undertow flows strongly back again. A horizontal shaft Savonius' rotor, nearly submerged, is driven in the same direction whichever way the flow might be. Such devices are clearly best installed where the tidal range is small. The plant described was at Monaco and it was used for pumping a sea water supply up to a marine laboratory. The Savonius rotor will, in the author's experience, generally start from any angular position, but it starts much better when an open wing faces the flow. Hence if several Savonius rotors are fixed on the same shaft, each with an angular displacement relative to the others, then at least one is always in a favourable starting position. This kind of displacement was used by Savonius in his wave application.

Savonius' work lives on, not so much as a windmill, for its capacity for steady work is very low, but as a current meter to measure the speed of the water deep in the ocean. Water is about eight hundred times more dense than air and so, even though ocean current speeds may be low, the forces are considerable.

7 Control

Most power plants are devices which receive a rather crude form of energy, moving air or water, or the heat from a furnace. Their function is to transform some of it into power available on a rotating shaft. This is the first step. The energy available from the rotating shaft can then be applied to generate electricity, to saw wood, or to pump water, or to do one of the many tasks which would otherwise, if done at all, be done by hand. It is only rarely that the task is of a steady and continuous nature. Commonly the demand on shaft power can vary quite markedly and suddenly as when a saw bites into a log, or when a number of lamps are switched on to a dynamo. On a national scale an electrical power system can receive sudden shocks as a by-product of television. On a generally quiet evening, a few minutes before the end of a popular programme, a million householders might decide that it is time for a cup of tea. All together they switch on their kettles and cookers. Equally important is sudden loss of load. This must be promptly signalled to the input end of the power plant, otherwise energy could accumulate and destroy it.

The constant balance of inflow and outflow of energy is well illustrated by the motor car where the supply of energy, that is of petrol vapour and air, is constantly varied, through the mechanism of the accelerator pedal, to suit the constantly changing demands of load, speed, acceleration and gradient.

In common with other engines, wind devices have to cope with varying load demands, but unique among the prime movers they have to cope with varying input as well. It is as though, in a motor car, a mischievous and unpredictable demon had an overriding access to the accelerator pedal quite independently of what the driver wanted to do.

There are thus two quite independent aspects of windmill control. The first concerns varying wind speed and the second varying output load.

When the wind speed is falling little can be done for we cannot turn the wind up again. Short term help could come from a flywheel or from an electric battery. When the wind speed is rising, at first energy is there to be used and then it is there in such abundance that some of the excess must be diverted before it does any harm.

The drainage mills of the Netherlands are still probably the biggest wind machines ever built to work reliably year in year out. The problems of their control can serve to illustrate various aspects of the subject. These mills normally worked a scoop wheel, that is, a waterwheel reversed to act as a pump. Initially the power might be sufficient to turn the wheel gently but not sufficient to force open the guard doors which prevent back flow from the upper level. As the wind speed increased the doors would be forced open and pumping would commence. Since the height of the lift would remain sensibly constant the flow, and hence the power used, would vary as the speed. However, if the speed increased much the water would not be able to flow into the wheel fast enough to fill the spaces between the blades. Shortly the paddle would beat only on air-water mixture. The load demand then could not increase as fast as the power available and the excess power would be available for acceleration. This could continue to destruction unless the windmill driver was able to apply sufficient braking or to turn the mill out of the wind. The general picture is of output load controlling the speed until the wind speed gets so high that some other mechanism, more or less an emergency action, must be brought in to maintain control.

Control by load

In the simple case of a mill in a steady wind (or indeed of almost any other prime mover under a steady energy supply) the power-speed curve must start from zero, for at zero speed, even if there is a torque, there is no power. As the speed is allowed to increase the power will increase also until it reaches a peak. Thereafter the power too will fall until it reaches zero at runaway, that is if the machine has not disintegrated in the meantime. The peak power corresponds to the design point where all velocities and angles are compatible. The falling part of the curve after the peak suggests that more and more of the power of the machine is being used up as friction in the machine itself. The process plotted on a graph gives the "power available" curve (V_1), Fig. 7.3.

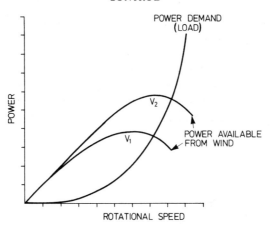

POWER DEMAND
(LOAD)

POWER

POWER AVAILABLE
FROM WIND

V_2

V_1

ROTATIONAL SPEED

Fig. 7.3   Matching the mill to the load

A reasonably small load approximated by battery charging and by some kinds of pumping also has a characteristic variation with speed. This is a "power demand" curve. Stable working is possible when the two curves intersect to the right of the peak. If the wind drops a little the consequent fall in speed means a fall in load also so that the speed does not drop too much. A rise in wind is met by a rise in resistance and is thereby controlled. A marked rise in wind speed leads to a higher "power available" curve (V_2). This might still intersect the "power demand" curve at a higher but still acceptable level. However, in the real world, ultimately the power available gets too high to be absorbed by any reasonable increase in load and this form of natural control must be supplemented if the mill is to be kept under control.

The power available can be prevented from increasing too much by reducing the effectiveness of the wind wheel and the power demand can be increased by the application of a brake. Sometimes both methods, diversion and dissipation of energy, are used. When there is an excess of energy in the approaching wind some of it must be either diverted, avoided or dissipated. It is not generally practicable to divert a windstream, but some of it can be avoided by turning the plane of rotation out of the wind. Exceptionally windwheels have been made which change in shape under the stress of the wind so that the area they present to the approaching flow is diminished. Thus a disk may turn into a cone (rather like a collapsing umbrella) or, in the case of Musgrove's vertical-axis machine, a cylindrical surface distorts into a conical one which becomes progressively flatter as the speed increases. Dissipation of energy can occur if an aerodynamic surface changes to one of lower efficiency and greater turbulent loss, or if a brake is applied. In this latter case there is an energy loss in the brake itself and there may also be increased aerodynamic loss in a wheel forced to run at a less than optimum tip speed ratio. Often the elements of avoidance and dissipation come in together. However, to bring any of these elements of control into play, either automatically or manually, the wind velocity likely to bring in the need of control must be sensed before action is initiated.

The sensing of change

The beginning of control is the sensing of change. If there is a human link in the chain of information forewarning of impending change can be a useful stimulus to vigilance. The weather services aid air and marine navigation in this way and are also of use in agriculture and in river control. When there is information of a storm in the mountains a lock keeper might lower his levels to accommodate the spate which will surely come. A windmill driver cannot have such exact foreknowledge. A forecast gale may or may not materialise, but he may see a cloud which suggests the imminence of a squall in time to take precautions. If there is no windmill driver but a mechanical device, it is not until the storm has begun that it can be sensed and control action initiated.

Changes in wind energy are made manifest by changes in its speed. This can be sensed by some form of anemometer. One of the buildings associated with a Dutch windmill often has a little windmill mounted on one of the gable ends. This could serve a practical as well as a decorative purpose. A glance at it could give the windmill driver an indication of the appropriate spread of canvas, just as the spinning cups of the anemometer outside a yacht club can give useful information to the initiated without the information ever being translated into a figure.

When automatic action is required it may be better to sense the force which the wind is exerting on a surface, rather than its velocity. When the wind is high this force may be of magnitude adequate to take controlling action. The force used may be either a lift or a drag or it may be the turning moment exerted by the flow on a sensing vane. The force may itself be used to turn the windwheel out of the wind or it may set some servo-mechanism into motion to achieve the same object.

Control by wind pressure

Control based on wind pressure protects only against overspeed due to excessive wind and it tacitly assumes that the windmill's load, generally pumping but sometimes battery charging, cannot be uncontrollably shed. However, overspeed can also arise from loss of load even in a moderate wind particularly in high-speed aerofoil machines. Control must then be exercised by sensing the rotational speed of the shaft.

Apart from protection against overspeed, an industrial process may need a fairly steady speed in order to give a uniform product. This too would need to be based on a shaft speed signal.

Shaft speed, which is not necessarily closely related to wind speed, is commonly sensed either by centrifugal or by electro-magnetic effects, the former when a substantial control force is needed, the latter for instrumentation. Some traditional windmills used flyball governors to control the flow of grain to the stones, thus indirectly controlling the load on the mill. Exceptionally such governors were used to open or close the shutters on the sweeps. Commonly today when variable pitch aerofoil blades are used, they may be feathered by the forces, or the moments, from the blades themselves, though commonest of all are centrifugal brakes.

Examples of wind-pressure control

1. A method commonly applied is to use the whole area of the windwheel disk as the sensing area. Even a machine with two narrow blades is quite solid to the wind when it is running at high speed. The windwheel axis itself is offset from the vertical pivot and is held in its normal running position by a spring. When the wind force exceeds a predetermined value the wheel tends to be blown out of the wind (Fig. 7.1). (A manual control may also be fitted which allows the wheel to be pulled parallel to the tail vane and so shut down completely.)

2. A rather more elegant method, which avoids the use of a spring, depends on the torque exerted on a control surface (Fig. 7.2). The main tail vane A is able to rotate about its fore and aft axis. This is

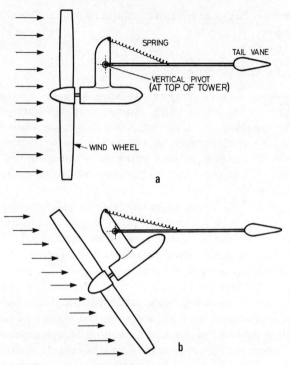

Fig. 7.1 Eccentric head control system (plan views)

connected to a control vane B through bevel gears. Normally the tail vane lies in the vertical plane and the control vane in the nearly (but not quite) horizontal one. When the wind is high the control vane, because of the moment acting on it, swings into a vertical plane where it acts as a steering rudder. At the same time the normal tail vane is turned horizontally, where it has no steering effect. The machine now swings out of the wind. Small machines using this kind of control are to be found in South Holland, where they gently urge the flow along drainage ditches towards the peripheral dyke. The principle employed uses the theory of the centre of pressure. Long before this principle was fully appreciated the method was used, but a very small third control vane was attached to vane B. This was initially in a vertical plane and its function was to initiate motion.

A recent British design for automatic control combines the effect of coning with blade angle by means of a simple and

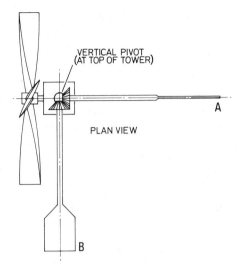

Fig. 7.2 Tilting-vane control system

carefully angled hinge. The disturbing force is the wind pressure. Restoring forces are due to centrifugal effects augmented by a spring. The device aims at high starting-torque, controlled running speed, and notably provides for total gust protection by full feathering, with automatic restarting. Field tests of the prototype have been encouraging.

Self-stalling blades

The outer extremity of a blade is the more important for it is here that most of the work is done. Even if a blade was so formed that the force intensity on it was constant throughout its length, the fact that power is force multiplied by velocity, means that the tip region is the more significant. Reasonably then, the angle of the blade should be such that the tip works under ideal conditions at the lowest wind speed contemplated. If a tip speed ratio of six to one is taken as an example, then this leads to a relative velocity angle of about nine degrees. The pitch angle must be about two degrees less than this, that is about seven degrees, to give the optimum lift/drag ratio. If the blade is made without twist, as we work inward along the blade, the relative velocity angle and the angle of attack both increase. At one-third radius the relative velocity angle is twenty-six and a half degrees and the angle of attack has risen to nineteen and a half degrees. The aerofoil is now stalled and is contributing little or nothing to the output.

In a sudden gust of wind the windmill cannot at once respond if it has a considerable flywheel effect and if a steady load is imposed upon it. Hence the stall will spread outwards towards the tip in a way which tends towards regulation. This kind of control has been carried further on machines synchronised into A.C. mains in such a way as to prevent any rise in rotational speed. The power can no longer rise according to the cube law since the higher the wind the lower becomes the efficiency of the mill. It is important to realise the limitations of this method. If the wheel became separated from the mains, i.e. its load, by a fault in the power line then the mill would at once come out of stall towards overspeed. Overriding emergency brakes are essential whenever this method is used.

Brakes

A mechanical brake of some form is necessary to hold the mill at rest when it is not in use or when it is being repaired or adjusted. The Aegean mills achieve this by use of a forked stick as a sprag, the fork-end engaged with a sail arm and the other sticking into the ground. The North-Western mills had a band brake; wooden shoes working on a wooden wheel for this purpose. Its use (other than as a parking brake) was fraught with peril of fire. A marked improvement for slowing down, but not for parking, was the introduction of the air brake. Air brakes, surfaces which may be turned normal to the wind were introduced into windmill work in the last century by a millwright named Catchpole. They have many advantages. They supply the control in the place where the force is generated, that is at the tips of the sails, and their effect is greatest when the speed is highest. Aircraft, from jet planes down to gliders, normally use air-brakes when coming in to land. Certain specialised types of competition motor cars do so at the end of a run. Almost by definition air-brakes are adequately cooled and there is nothing to wear away. High-performance windmills often have air brakes as an overspeed emergency method of control. These may be plates which are normally fixed into the wing but which can, when required, swing out at right angles to the wing. Some of the successful Danish machines could rotate the tip of the blade through ninety degrees. Whilst air brakes on the sails have advantages they can be a source of danger on high-speed machines. Their action might be such as to induce metal fatigue or under icing conditions they may work unevenly. Anything which puts a machine out of balance by even a small amount can generate damaging centrifugal forces. A preferable method might be to incorporate air brakes on a smaller auxiliary wheel mounted firmly between the main bearings of the windshaft. Braking will not present problems if ever we start building substantial windmills again.

The windshaft loading conditions of a large European mill are just about the same as those of the rear axle of a heavy lorry grinding up a steep hill in bottom gear. Appropriate technology for this kind of application is now very well in hand. Engineers could build really good windmills now if the politicians and the accountants gave the all clear. But they could meet with opposition from amenity societies. It is interesting to speculate which action would bring the strongest protest, a proposal to build a new windmill in a necessarily exposed and prominent place or a proposal to demolish an old one (which had been up so long as to be accepted), in a corresponding place.

The automatic governing and regulation of a windmill is difficult but it can be done and it has been done in a few special cases where alternative sources of power would be more expensive.

Windmills have been used on buoys moored far out in the ocean, the power being used for the collection and transmission of oceanographic and weather data. They also work in deserted places as an aid to radio and telephone communication and they are used to work navigation lights on isolated hazards.

These applications are all small and they have been used in places where other devices would have been too expensive to maintain.

When size is increased the structural and control problems of the aerodynamic surface become more difficult. There is probably an optimum size, which may not be very large.

There is also a difference in complexity between an isolated self-contained unit and one which feeds into a power network.

It is interesting to note that the sail mill, the simplest and cheapest of all, has the inherent property of regulation.

Windseeking and the tail-vane

The natural wind even while maintaining a steady average direction is normally subject to frequent minor changes. Experience shows that windmills work best when they are firmly held facing the mean direction of the wind. Attempts to follow minor changes are counterproductive as regards power production and mechanically they are undesirable.

Small windmills mounted upstream of a mast are commonly pointed in to the wind by means of a tail vane, and this may be combined with more general methods of control as described earlier in this chapter. Despite its apparent simplicity a windwheel tail vane combination of this kind involves an inherent directional instability. This instability can be relatively small and its effects can be suppressed, as they are in the many thousands of successful windmills in daily use. However, radical departure from proved practice may have unexpected results. Further complications can arise if the tail vane is mounted in the disturbed wake downstream of the wheel and the mast, where the flow velocity is both reduced and disturbed.

From the results of his laboratory experiments the author concludes that the following are the desirable factors tending toward directional stability (but as some of them may conflict with other desirable requirements, compromise is often required).

1. Short overhang of the windwheel in front of the mast.
2. Long overhang of the tail vane behind the mast. In the author's view this should at least be equal to the diameter of the wheel.
3. Mounting the tail vane in the undisturbed flow above the windwheel.
4. Providing means of damping motion around the vertical axis. This may well be incorporated in the design of the thrust bearing which takes the weight of the swinging assembly.

Downstream windwheel

A windwheel mounted downstream of the mast has an apparent simplicity in that no tail vane is required. However, the elimination of the tail vane also eliminates one means of control. Downstream windwheels have mostly been used on small aerofoil machines which could be supported on a relatively slender mast. Control has generally been supplemented by use of a side vane which can be tilted to swing the machine out of the wind, or by a brake, either a brake operated manually from the ground or by an automatic centrifugal device. The downstream windwheel may be a necessary feature of mills which govern by allowing the blades to cone inwards, though this is associated with mast clearance rather than with windseeking, the latter function being taken over by a vane-controlled servo motor.

8 Wind and power measurement and mill performance testing

The Beaufort Scale of Wind Force

Wind studies and the shipping forecasts still, in some degree, perpetuate the memory of Francis Beaufort. Beaufort was born in Ireland and he rose to be Hydrographer to the Admiralty, an Admiral of the British Navy, and a Fellow of the Royal Society. As a young officer in the British Navy he sustained nineteen wounds in one particular action and while recovering he set out a line of semaphore telegraphs from Dublin to Galway. This could, it is reported, transmit a message and get a reply in eight minutes. Subsequently other such telegraphs were established, such as London to Portsmouth and Holyhead to Liverpool.

In 1805 Beaufort was back at sea in command of the frigate "Woolwich" and it was then that he devised the wind force scale which bears his name. Beaufort's object was to devise a concise and repeatable method of reporting wind conditions in the ship's log. Such information was needed for Admiralty records, sometimes for investigations or for courts martial subsequent to actions at sea as well as for the accumulation of geographic knowledge. Beaufort's scale was based entirely on the behaviour of a frigate of 1805. In effect the whole frigate was his dynamometer. Note well that Beaufort's is a force scale and not a speed measurement.

Force 1 was that which gave her steering way but no more. Force 4 was that which with all sails set could push her speed up to 6 knots in calm water when she was sailing on her best course, then described as "clean full". Force 7 considers the ship "in chase" which perhaps means that she is carrying as much sail (and perhaps a little more) than is really prudent. Her royals have come down and her topgallants are single-reefed. At Force 11 the phrase "in chase" has been omitted. Perhaps emphasis has changed from combat to survival. It is the condition when, to attempt to keep some control, she is carrying her storm trysails only. Force 12 is "that which no canvas can withstand". All the definitions are based on the behaviour of the ship and not on the appearance of the sea.

These numbers were exact and without ambiguity to the men who had to use them. As the years passed and frigates became less common the scale was redefined. Thus Force 5 became that "at which smacks shorten sail". It still is! The condition "in chase" corresponds perhaps to racing, in that a yacht on a cruise, with the wind gusting five, might for more comfort shorten sail, but on a race her master might tend, with greater vigilance, to hang on a little longer. Redefined, the numbers are intelligible to seamen today when frigates and people who know how to handle them have long ceased to exist. Some recent versions of the Beaufort scale have changed from the behaviour of the ship to the appearance of the sea.[12]

Shipping losses in the Crimean War due to unexpected storms were worse than those due to combat. This was the stimulus for a more detailed study of the weather. The barometer, the cup anemometer and the electric telegraph then provided the preconditions for the construction of synoptic weather charts. Hindsight studies of the passage of storms across a region, with their corresponding changes of pressure and shifts of wind, began to suggest means of forecasting. Such studies could be supplemented by records of weather conditions which had been logged by ships at sea. Anemometers on land gave windspeed values, in units such as knots, miles per hour, or metres per second. The figures in the ships' logs gave wind force numbers. Such was the strength of maritime tradition that the land stations had to recode their data into force numbers. For about a hundred years attempts were made to link Beaufort numbers, based on a seaman's "feel", with speeds recorded on an anemometer. There was never complete agreement but often complete confusion. Competing scales of equivalents might differ by as much as a hundred per cent.

In the latest and most accepted version of the Beaufort scale, Force 3 is given a mean speed of 4.4 m/sec (10 mph) and Force 6 is given one of 12.6 m/sec (28 mph). This spans the practical small windmill range. Force 3 is that which begins to give some useful output. Through Forces 4 and 5 the mill is working so well that the driver wonders why anyone should be interested in any other source of power. At Force 6 the windmill driver is beginning to get uneasy and planning to call a halt. As he does so he can remember Captain Beaufort in 1805 ordering a single reef in the topgallants of H.M.S. "Woolwich" (in chase).

Beaufort scale numbers 3, 4 and 5, which are described as gentle, moderate, and fresh, respectively, cover the practical range of speeds for windmills, that is, from 3.6 m/sec (8 mph) to 10.8 m/sec (24 mph). With only three steps covering a power range in the ratio one to twenty-seven (i.e. depending on the cube of the wind speed) the scale is clearly too coarse for practical use in windmill testing.

The pitot tube

Aerodynamicists regard the pitot-static tube as an absolute measure of speed (see Fig. 8.1). The pressure p_p recorded when an open tube points into the stream is greater than the pressure p_s recorded at the

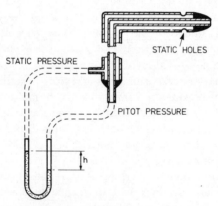

STATIC HOLES

STATIC PRESSURE

PITOT PRESSURE

h

Fig. 8.1   The pitot-static tube

static holes perpendicular to the stream. The difference is related to the wind speed by the equation $p_p - p_s = \frac{1}{2}\rho V^2$. The Greek letter ρ (rho) signifies the density of the air. The pitot-static pressure is given by the equation $p_p - p_s = \rho_1 gh$, where ρ_1 is the density of the liquid in the gauge and h is the height of the liquid column.

The pressure difference $p_p - p_s$ for light winds is so low that precise manometers are needed to measure it. This effectively restricts the instrument to the laboratory when the speeds are low, but it becomes a completely practical field instrument at higher speeds. Indeed it is the speed instrument on which air pilots depend.

The anemometer

When direct methods are not practical, as they are not in the measurement of moderate wind speeds, then inferential methods must be used which in turn need a laboratory calibration. A well known inferential device is the rotating cup anemometer. This can be used in two ways. (1) The total number of revolutions can be counted over a period of time. These should be proportional to the "miles of wind" passing in that time. A mechanical gear in the anemometer can either work a counter directly, or it can trip an electric switch which in turn operates a distant solenoid drive counter. (2) The instantaneous speed of the anemometer can be inferred and by calibration this can be related to the instantaneous wind speed. Commonly, a small permanent magnet alternator is driven by the anemometer. Provided that only a small electric current is drawn then the potential difference across the alternator terminals should be proportional to the speed. This potential difference can be displayed on an electric meter, which is calibrated to read wind speed on its dial. The very small cup anemometers now commonly fitted to sailing vessels may depend on an electrical pulse originating from the movement of a permanent magnet near a magnetic reed switch in the anemometer body. The pulse rate can be modified by electronic devices to give a display on a meter in the cockpit.

A home-made anemometer can be contemplated. Its viability would depend on low-friction bearings remaining consistent when working in an exposed environment. (Note that if an attempt is made to calibrate such an anemometer by mounting it on a motor car, the distortion of the streamline caused by the passage of the vehicle itself should be borne in mind. If the anemometer is mounted on a mast as high again as the car, then its reading should be at least as good as that of the car's speedometer. All that is then needed is a private road, a windless day and a uniform speed.)

Hand-held anemometers for the use of mariners can be obtained from ship chandlers. One type picks up the pitot-static pressure of the wind. This displaces a disk which can rise up and down inside a tapered and transparent container of circular cross-section. The position of the disk corresponds to the wind speed and this can be read off directly on a scale. The validity of the instrument depends on precise and consistent manufacture. The author has found them to be remarkably good when put to proof in a wind tunnel.

Performance testing of windmills

Performance testing is resorted to when a machine is too complex for analysis. This applies to nearly all of those involving fluid motion. There are generally many variables and the key to the understanding of experimental results is to arrange for only one of these to be changed at a time. The experimenter must decide on the property which is of most concern. This is the dependent variable. Now all the things which could separately influence this must be considered. These are the independent variables. If an experiment is now arranged in which only one independent variable is changed (all the others being held constant), a set of systematic results can be obtained which are characteristic of the machine. When these are plotted on a graph the result is called a characteristic curve. It might seem that to obtain the whole picture each of the independent variables would have to be changed, one at a time. This could lead to a daunting experimental programme. Fortunately mathematicians have devised ways of grouping the variables so that the experiments needed to cover all the possibilities can be greatly reduced. These methods are gratefully accepted and used by engineers. Without them engineering science would long ago have been overwhelmed and sunk beneath accumulating data.

A windmill laboratory

The facility to be described as a windmill laboratory was developed from lecture demonstrations which the author devised over a number of years. It is not in any way to be compared with a wind tunnel and it is not suitable for instrument calibration, but it does enable windpower experiments to be made which are repeatable and are not unrealistic as to size or performance with results which may be obtained in the field. Furthermore, from an educational point of view these air-machine experiments can complement the work of a hydraulics laboratory.

The artificial wind derives from a small ex-aeroplane propeller of approximately 1.75 m (5¾ ft) diameter and 1 m (3⅓ ft) pitch. This is mounted on a stand in a large open room and it can be electrically driven at speeds up to 400 rpm. At this speed the wind velocity produced at the test position 1.5 m (5 ft) in front of the airscrew is about 5.0 m/sec (11 mph). The propeller is driven from an ex-aircraft dynamo, used as a variable-speed electric motor, and its power requirement is a little less than 1 kW.

During a test, as a safety measure, both the propeller and the wind-mill under test are surrounded by wire netting screens 2 m (6½ ft) high. These are firmly clamped in position.

The experimental windmills used vary in diameter from 0.75 m to 1.5 m (2¼ ft to 5 ft) and are thus in the range of practical machines capable of doing useful work. (See Fig. 10.1, page 100.)

Units of measurement

How often do practising engineers, when confronted with drawings or calculations in unorthodox form, echo that cry from the heart recorded in Proverbs, "Divers weights, and divers measures, both of them are alike an abomination to the Lord". The author is told that the text is of Egyptian origin, perhaps first uttered by an exasperated site engineer on a pyramid project. Measurement systems can be rational provided everyone sticks to them, but often they do not. James Watt chose as his unit of power the brewer's dray horse, but he adjusted his experimental results by fifty per cent. He wanted to make more than sure that his steam horses could work harder than real ones, in case an ungrateful customer should endeavour to put an engine to proof.

Of late years there has been a possibility of rational usage in a refined version of the metric system, in what is called the S.I. (Système Internationale). This is based on the kilogram mass, the metre length, the second of time, and the Celsius (Centigrade) degree. James Watt is commemorated in the unit of power, called a watt, but it is no longer based on what a brewer's dray horse can do. If this system could be wholeheartedly accepted education could be simplified and understanding increased. Unfortunately there is no co-operation from the great energy industries which impinge on everyone. The electrical industry charges by the kilowatt-hour. Gas perpetuates the Fahrenheit temperature scale in the therm. These differences make rational comparison of costs more difficult than they need be.

Shaft-speed measurement

Rotational speed is measured by counting the number of revolutions made by a shaft over a measured period of time. This method is absolute and it is the one to which other methods must be referred. However, when steady conditions are assured inferential methods may be used and they can be very convenient. For instance, if a small permanent magnet generator is driven directly by the test windmill it should (provided only a minute current is drawn from it) give a voltage across its terminals which depends only on the speed of rotation. This voltage can be displayed on a suitable meter which in turn can be calibrated (by counting and timing methods) to read directly in rotational speed. Such an instrument which gives a direct reading of rotational speed is called a tachometer.

When working in the field the author has found that revolution-counting over a considerable period gives more consistent results than those derived from a tachometer. The time has been in the region of ten minutes to half an hour. When a remote indication is needed the windshaft can be made to operate a micro-switch which in turn drives an electro-magnetic counter at the test station.

Measurement of power and the dynamometer

Devices used for the measurement of force or torque are called dynamometers. These can take many different forms, ranging from very simple to complicated devices. One of the simplest forms is a type of brake which absorbs all the power being developed and which is arranged for the measurement of the relevant forces. The measurement of power on such devices has given rise to the term "brake horse power". Other dynamometers do not absorb the power but enable it to be measured while it is being transmitted from one device to another. These are transmission dynamometers. Yet another measures the torque which must be applied to prevent the driven machine rotating bodily. This is a torque reaction dynamometer.

When a machine is driven by a rotating shaft, the whole machine tends to rotate with the shaft. This is prevented by some holding device which must apply a torque equal and opposite to that being fed in. Sometimes it may be practical to measure this reaction torque, in which case it is a convenient step towards the measurement of the power.

This type of measurement is particularly easy when the driven

machine is a dynamo with a circular casing. The casing can readily be
supported by three sets of rollers (which may well be ballraces) (Fig.
8.2 at C). The rotation of the casing can be prevented by forces
applied to the end of a radial arm of length R_1. The movement of the
arm is restricted by robust stops.

The use of both a spring balance and a weight tends toward greater
stability.

A simpler type of brake consists of a rope of radius r passing
round a wheel mounted on the windmill shaft (Fig. 8.2, B). Tensions
are produced in the rope: at one end by the weight of the mass
M (i.e. Mg) and at the other by the pull of a spring balance m (i.e.
force = mg). The rope commonly makes either half a turn round the
wheel (as illustrated), or if more friction is required the rope can
make a whole turn round the wheel, in which case the guide pulley
is not required.

The torque T applied by the rope brake is given by:

$$T = (M - m)(R_2 + r)g.$$

If the S.I. system is used M and m are measured in kilograms, R and r
in metres; g has a value of 9.81 m/sec^2.

Power is torque multiplied by angular velocity $2\pi N$ where N is the
rotational speed in revolutions per second.

The power in watts then is:

$$(M - m)(R_2 + r)g\, 2\pi N$$

The power in horse power is:

$$\frac{Mg(R_2 + r)\, 2\pi N}{550 \times 32.2}$$

In this case M is in lb, g is 32.2 ft/sec^2, R_2 and r in feet, N is in rev/sec.

NOTE—The pure numbers 550 and 32.2 arise from the arbitrary
nature of the horse-power. The 32.2 is not the same as g even though
it may be of the same numerical value. One is an acceleration and the
other a pure number. The number 550 arises from James Watt's
definition of a horse power.

Measurement of power on the model scale

The methods just described are not particularly suitable on the
extremely small scale. Uneven friction and snatching of the brake or

friction of the rollers becomes more significant as the scale is reduced. However, an alternative method is available which is particularly valuable in an educational context, since it brings all measurements down to the basic mass, length, and time.

This method consists in the direct lifting of a weight by the windmill. Conveniently a load may be suspended by a thread which is allowed to wind up on the windmill shaft.

If the load M is lifted through a height z in t seconds, the power is Mgz/t. If M is in Kg, g in m/sec², z in m, and t in sec, the power is in watts.

NOTE—The distance from the ground to a must be enough to allow the windmill to acquire a steady speed. The distance from b to the shaft must be sufficient to allow the windmill to be stopped easily before the weight reaches the top.

Electrical power

The electrical power developed by a direct-current dynamo is given by watts = volts × amps, where the volts and the amps may be measured by separate meters. This is always less than the mechanical power fed into the shaft. Losses arise from bearing friction and windage as well as from electrical and magnetic effects within the machine. Small electrical machines are not always very efficient. One which delivers 70 per cent of the shaft power as electrical output is quite good.

Performance testing: practical points

The dangers inherent in moving machinery must not be overlooked. Even a low power machine may have a lot of energy stored, as in a flywheel, and it must be treated with respect.

The simple friction brake seen in Fig. 8.2B can be taken as a case in point. The load Mg can be quite large even in a small machine of low speed. If the rope and the pulley are not compatible jamming might occur, with the result that the large load might be hoisted up in the air and whirled around the shaft. Twenty-two kg (48 lb) of iron whirling round on the end of a rope twenty feet up in the air has a great potential for injury. The author has known this to happen when precautions which he had requested were disregarded. Fortunately nobody was hurt, but somebody might have been.

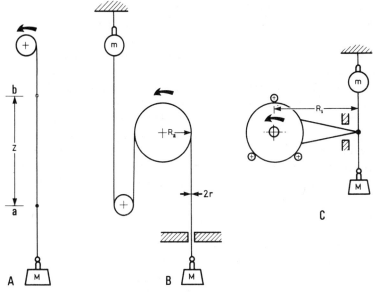

Fig. 8.2   Types of dynamometer

Points to observe

1. Prevent jamming at the source. The pulley flanges should be such
   that the rope cannot jam mechanically. Ideally, a wide flat pulley
   with flanges can be used, or if a Vee pulley is used its width at the
   bottom of the "V" should be much bigger than the rope
   diameter.
2. Use a well-oiled natural fibre rope. The author has never had any
   success with synthetic ropes used as brakes.
3. Always build in an adequate mechanical safeguard which can
   restrain the weight even if the rope jams. It should be made quite
   impossible for the weight to lift more than a few inches.
4. Instead of using ordinary iron weights, weights can be made as
   required by using plastic bottles filled with water. The weight of
   the bottles must be known and, provided water is available at the
   site, a large range of weights can be built up by use of a measuring
   can. The saving in labour can be significant!

Testing in the laboratory can sometimes be straightforward but in the
field this is rarely so. The wind seldom keeps steady for more than a few

seconds at a time, with the result that the output of the windmill is un-
steady also. In addition, instruments must necessarily lag in their
readings. When a figure is read on an anemometer dial it is, at the best,
the wind speed a few seconds ago. The same is true when the torque
developed by the windmill is recorded. If the wind is momentarily
slowing down, the brake reading remains high, for both the input
torque and the torque being given up by the flywheel effect of the mill are
operating. Again an underestimate might be made if the wheel is being
accelerated. It is not surprising that an experimenter may sometimes
find it difficult to get repeat results of his own work, let alone of other
people's. There is always a big scatter from results taken in the field and
this must be borne in mind when comparing different people's work.

The most repeatable results are obtained if, on a day of fairly
steady wind, readings are summarised over a considerable period of
time, say half an hour to one hour. The anemometer in such a case
should be one which can give an integrated "miles of wind" over the
whole period, and if possible the work done should be accumulated
also. This is easy if it is a matter of pumping water into a reservoir,
but not practicable if the work is being absorbed into a brake.

Testing technique: observed and derived results

An illustrative case is a small windmill set to face a constant wind in a
laboratory. The primary interest is in the maximum power which it
can develop in this wind and in the rotational speed at which this
develops. The information from such a test might be necessary to
decide on an appropriate gear ratio between windmill and dynamo.

The use of plotting

Although it is power which is being sought, it is weights and speeds
which have to be observed. In this case the brake weight and the shaft
speed figures are the observed results. Subsequently power is to be
derived from them, but first a guide as to procedure.

Observed results are not exact. Spring balance needles wobble
about and successive speed readings may not be consistent. If the
brake readings M and m (Fig. 8.2) are plotted against the rotational
speed N they will generally suggest a smooth curve (Figs. 8.3 and 8.4)

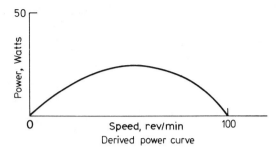

Derived power curve

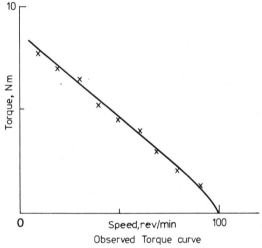

Observed Torque curve

Fig. 8.3   Sail mill, 1·5 m (5 ft) dia.: Torque/power curves

which can be drawn through them, but not necessarily touching any of them. There is a lot of experience to suggest that this smooth curve represents something nearer to the truth than do the isolated points. Departure from the curve is generally an indication of experimental error. Hence, if this can be accepted, when once the smooth curve has been drawn the original points should be forgotten and further information must be taken from the curve. Of course the experimenter must be aware, some phenomena may take place in jumps and jerks, in which case smoothing could hide them. On the whole, however, in these simple experiments discontinuities are not likely to happen.

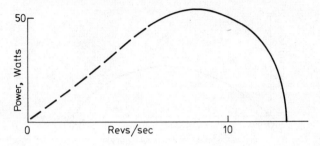

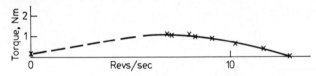

Fig. 8.4   High-speed aerofoil mill, 1·5 m (5 ft) dia.: Torque/power curves

The selection of scale used in plotting a graph is important or mis-leading results may be obtained. As a guide, the smallest divisions on the graph paper should correspond to the smallest unit to which an observation can be made. Even more important in scientific work is the use of true origins. Let both the graph scales start from zero. False origins, unless very clearly understood, can be misleading.

When a small windmill is tested in a laboratory with all the variables under control it can be made strong enough to withstand runaway conditions under the laboratory wind. This makes observation of the complete characteristic possible.

The form of the torque curve depends on the type of windmill being tested. The example of Fig. 8.3 is taken from a sail-mill rigged for maximum starting torque, whilst that of Fig. 8.4 is from a high-speed aerofoil of the same disk diameter.

The curves are taken by first letting the windwheel achieve runaway speed (under the controlled laboratory wind). The speed is noted and then increments of weight are added to the brake, the speed being noted in each case until the windwheel comes to rest. This is a gradual process with the sail mill but it is quite abrupt in the case of the high-speed aerofoil. Soon after the maximum torque the foil goes into stall. It is not generally practicable to observe the unstable (dotted) part of the torque curve.

Whatever the form of the torque-speed curve, the power-speed curve must start from zero when the machine is at rest and return to zero at runaway. Even if there is a large torque at rest, if there is no movement there is no work being done. Similarly at runaway, even though there is a lot of movement, if there is no torque then there can be no work done. The power-speed curve must then have a maximum value between the zero values. The location of the speed at which this maximum occurs is an object of the test. The speed at which the maximum power occurs generally suggests the required gear ratio between the windwheel and the driven machine.

An application of similarity

When a set of characteristic curves have been plotted for one machine at one wind speed, the methods of similarity enable the characteristics of larger or smaller machines in other wind speeds to be predicted. If very small and very large velocities are excluded then the application is particularly simple. No proof is attempted but it can be accepted that the method is well established.

The relevant result is that at any one tip-speed ratio the power of any one of a series of geometrically similar windmills would depend on the square of the diameter and the cube of the wind speed.

As an example (working in metric units only for precision and clarity) assume that the power of a windmill of 1 m diameter is 20 watts at 200 rpm in a wind of 5 m/sec. What would be the power of a geometrically similar windmill of 4 m diameter in a wind speed of 10 m/sec?

If suffix 1 refers to the small windmill and 2 to the large one, the above written statement can be expressed in symbols:

$$\frac{P_1}{D_1^2 V_1^3} = \frac{P_2}{D_2^2 V_2^3}$$

$$P_2 = P_1 \left[\frac{D_2}{D_1}\right]^2 \left[\frac{V_2}{V_1}\right]^3$$

$$= 20 \times 16 \times 8 = 2,560 \text{ watts}$$

P represents power, D represents diameter, and V represents wind speed.

The speed follows from the requirement of constant tip-speed ratio:

$$\frac{N_1 D_1}{V_1} = \frac{N_2 D_2}{V_2}$$

$$N_2 = N_1 \ \frac{D_1}{D_2} \frac{V_2}{V_1} = 200 \times \tfrac{1}{4} \times 2 = 100 \ \text{rpm}.$$

9 Reasonable expectations

Calculation of power available

The possible range of energy which a windmill might deliver is very easily calculated. The kinetic energy of a stream of wind of cross-sectional area A is given by the equation Energy $= \frac{1}{2}\rho AV^3$ where V is the velocity of the wind. The density ρ is not likely to differ much from 1.1 kg/m³. If the metre-kilogram-second (S.I.) system of units is used the power comes out directly in watts. If there is interest in the imperial system of units, then 746 watts is equal to one horse power.

Thus, for example, a wind-speed of 4.5 m/sec (10 mph) corresponds to a power of $\frac{1}{2} \times 1.1 \times 1 \times 4.5^3 = 50$ watts per m² (10⅜ sq ft) of area swept by the windmill. This is the energy of the wind. Accepted doctrine shows that at the very best an absolutely perfect windmill could only extract 59.3 per cent of this. In real life this could not be attained. In fact, a machine which extracted 50 per cent, or even say 30 per cent, of the wind energy would be a good one. Thus a machine which can deliver 15 watts per m² at the rather low wind speed of 4.5 m/sec (10 mph) is doing reasonably well.

The equation for the wind energy $E = \frac{1}{2}\rho AV^3$ enshrines the principle and abiding problems of wind power. Since ρ, the density, is what nature furnishes, it is always small compared with that which can exist in an engine or turbine and the values of wind speeds which must be used are small also. Long experience in rural England has shown that wind speeds from 4.5 m/sec (10 mph) to 11.25 m/sec (25 mph) are potentially useful. It is of little use to design a windmill that only gives useful power in an exceptional storm. Both these factors mean that a windmill is large for its power when compared with other machines. Also, whilst the windmill must be sufficiently light and free to develop power in the lowest useful wind speeds it must yet be strong enough to withstand the strongest winds which ever blow. Thus a machine which is giving useful power at a wind speed of 4.5 m/sec (10 mph) must withstand a storm of 45 m/sec (100 mph), when the forces are one hundred times as great and the energy has increased one thousand fold.

Six windwheels compared

The following results are based on laboratory experiments carried out by the author. The wind speed was 5 m/sec (11 mph).

Type	Watts/m²	Tip-speed ratio
High-speed aerofoil (Two-blade)	29.0	8.0
Low-speed aerofoil (Four-blade)	27.0	1.8
Sail mill rigged for maximum power	26.0	2.1
Sail mill rigged for maximum torque	14.0	0.8
Inclined flat boards (Two-blade)	5.0	1.5
Savonius rotor	1.5	1.0

A traditional Dutch windmill of 26 m (85 ft) diameter is reported to have achieved 55 horse power in a wind speed of 10 m/sec (22 mph) at a speed of 20 rpm. When scaled to the terms of the above list this is consistent with 10 watts/m² at a tip-speed ratio of 2.7. Knowledge of a machine's capabilities in a given wind speed is the first step. An estimate of the energy it can deliver over a whole year on a given site is the next, and a more important one.

The site and its effects

To have a reasonable chance of success wind power plant needs to have a free and undisturbed flow of air. Wind "shadows" downstream of an obstruction can extend for great distances. A single haystack 4.5 m high and 9 m (30 ft) long, on a prairie, is reported to make its influence felt for 300 m (1,000 ft) downstream. It is not for nothing that traditional mills were, whenever possible, built on open hills. In the absence of hills artificial mounds have been raised in the Netherlands. City mills, such as those which exist in Schiedam, stand very high indeed above the city. So it is as well to expect little more than a decorative mobile from a windmill built in a suburban garden. It is the violent turbulence and apparently strong wind that can be experienced in an urban area which gives a misleading impression of available energy. The author has experimented with a small axial flow windmill in such an area, where gusts and eddies made walking difficult, yet found that they signified nothing as regards the steady production of mechanical power. Given that a clear approach exists from the direction of the prevailing wind it may still be that for part of the year the power potential is dominated by a highly local wind system. As a yacht race timekeeper the author has, while spending long vigils on pierhead lookouts, been able to watch brisk inshore sailing whilst only a few miles away the offshore fleet for which he was waiting has been completely becalmed. Ideally the evidence drawn from personal experience of wind conditions on a particular site should be supplemented by confirmation which could come from a recording anemometer operating for a representative period of time, and working in conjunction with a pilot plant wind-mill of a size which can give a useful power output.

Guidance as to energy expectations on an inland site in Southern England can come from the comprehensive trials carried out by a research institute of the University of Oxford more than fifty years ago. Nine windmills operated the whole year round on a site near Harpenden, Hertfordshire, and the electrical energy which they delivered was cumulatively measured. The machines varied in design from American type windwheels to medium speed aerofoil machines and the size from 2.4 m (8 ft) to 9 m (29½ ft) diameter. The best of the aerofoil machines had properties which would be commendable today. The lowest and the highest annual outputs were 83.5 kW hr/m^2 and 199 kW hr/m^2, respectively. Both these came from multi-bladed sheet metal wheels at heights of 3.3 m (11 ft) and 16.5 m (54 ft) respectively. The best aerofoil machine gave an annual yield of 165 kW hr/m^2.[13]

A suggested interpretation is that a well-designed axial flow machine, at a height of 10 m (33 ft), should yield approximately 100 kW hr/m^2 each year on a good level inland site in England. Published wind information,[14] suggests that more could be expected on a hill or on the coast. Much more could be expected on an out-lying island.

In the author's view the significance of these figures is that when there is only a small amount of energy available then high-grade energy should be used for high-grade purposes, such as mechanical work, and (he is prepared to concede) for light. Small and intermit-tent supplies of low-temperature heat might be more safely, and more cheaply, acquired by building an effective furnace to use small quantities of otherwise waste fuel than by degrading wind energy. Of course there are exceptions such as when really quite small amounts of heat energy need to be applied in highly localised manner.

10 Construction of a pilot plant

Hazards and limitations

Anyone who sets out to build a windmill should not overlook the possible hazards. Accidents to bridges, television masts and cooling towers contribute to a continuing history of the underestimation of elemental forces, even by engineers with great resources behind them. For a tower to collapse is bad enough. For a high-speed windmill blade to come adrift is potentially worse. It is implicit in the literature and confirmed by the author's experience that a machine tends to deliver less power than had been hoped for and that its tendency to come to grief is greater than had been expected.

Influences other than wind pattern and intended duty are the skills and materials available to the builder and the degree of supervision that is given to the plant when it is at work. Non-technical factors can also come into account. Prominent among these are the attitude of neighbours and of authority, though these are not always unfavourable. No single answer can be given to such questions as, "What is the best kind of windmill to build?" Every reply must be hedged with compromise and qualification. It can be said that ability to survive quite extreme conditions is always an essential quality, as is a built-in tendency to self-regulation. Control should always take precedence over output. It can be taken as axiomatic that the alternative is a recipe for disaster.

A pilot-size sail machine

Whatever type of machine is eventually intended, the author considers that a small sail mill should be built and operated before the main project begins. Regarded as a pilot plant such a machine can give an assessment of the power potential of the site and it can also be a means of furthering the education of the builder. The sail mill is the cheapest and easiest of all to construct and if carefully adjusted it can, in the author's experience, give an output comparable to that of the aerofoil mill of the same diameter, but at a much lower speed. Such a mill should initially be under manual control and it should have a brake capable of bringing it to rest.

An eight-arm sail mill 1.5 to 2 m (5 to 6½ ft) diameter is convenient in that it can demonstrably do useful work and, considered as a model, it is large enough to be free from significant scale effects.

Advantages accrue from the use of a mobile rig for test purposes. No windseeking device is involved and the potentiality of an area as regards position and direction can be investigated in detail. In time of threatened storm the whole thing can be towed away to shelter, always provided that it is not too high. Attractive windmill test sites are necessarily open and exposed to the public view. Often they are in areas of high amenity. Permission for research work may be more readily granted when it is clear that no permanent structure is involved and that removal can be expected on completion of the tests.

The author has built several experimental test stands to the basic design of Fig. 10.1 for laboratory work and lecture demonstration. Exceptionally, and when adequately ballasted, they have been used out of doors for special investigations. The basic structure is of timber of 75 × 50 mm (3 in × 2 in) rectangular section secured by through-bolts with nuts and washers at every joint. Mass-produced

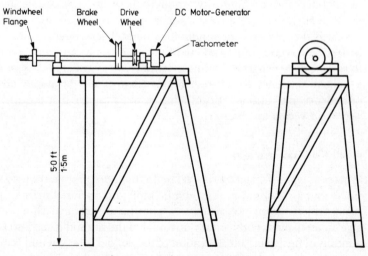

Fig. 10.1   Laboratory test stand

engineering components are available from specialist power transmission engineers, which can be brought together and incorporated into a wide range of machinery. Those relevant to a simple windmill application can include pulleys, sprockets, bearings, flanges, and shaft-mounted gearboxes. The components are made in a range of sizes based on shaft diameter.

General purpose mild steel shafting is available from the steel mills in either black or bright bar. Black refers to a coating of oxide on the surface which indicates that the final process in the steel mill was carried out at an elevated temperature. The term bright indicates cold treatment. Commercial products are necessarily made to a nominal size with some degree of tolerance as to dimensions. The tolerances for black bar may be positive or negative while those for bright bar are smaller and negative only. That is, bright bar is either very near to the nominal size or is slightly less. The standard fittings listed above are made very close to the nominal size and should thus fit readily on to the bright bar. Often these fittings and pulleys incorporate means whereby they can be attached to the shaft with the aid of very simple tools.

The basic windshaft for the prototype or demonstration mill requires a suitable length of bright bar and a pair of self-aligning ball bearings mounted in plummer blocks. Other fittings are a flange for the windwheel, a brake wheel, and some form of power take-off. 25 mm (1 in) is a suitable diameter for the windshaft of the smallest type of machine.

A "Vee" belt drive can be used off the windshaft of a 2 m (10 ft) machine, the type of belt and diameter of the pulleys being best chosen to be in accordance with the belt maker's recommendation and the required speed of the driven machine. For electrical generation two stages, with a countershaft, may be needed. For a 4 m (13 ft) diameter sail machine the torque is so high that a positive drive, gear or chain, is desirable for the first stage of a speed-increasing gear.

The projection of the windshaft upstream of the flange can be threaded so that alternative windwheels can be bolted firmly up to the flange and an extended nut can form an attachment for a bowsprit.

The author's preferred windwheel construction utilises eight sail arms radiating from a flat disk of marine plywood. This in turn is secured to the flange on the windshaft. The sail arms are of rectangular section where they are bolted on to the disk but rounded off to an elliptical section and given about a 10° pitch where they receive the sails. This helps the air flow to pass gently on to the sail surface. A 2 m (6½ ft) diameter mill of this type has a central disk 0.3 m (1 ft) in diameter and the arms are formed from hardwood 50 mm × 40 mm (2 in × 1⅝ in) section. (A mill of this kind with all the sail removed continued rotation in a gale, though this had not been intended.) Numerous ways of attaching the sails to the arms suggest themselves. Perhaps a sleeve sewn on the sail is as good a method as any in that the sail can be easily removed. One 4 m (13 ft) diameter mill has been built with luff grooves formed in the sail arms which receive a rope sewn along the edge of the sail. The general conclusion was that no appreciable benefit accrued. Larger structures, all dimensions being increased in proportion, have been built for outdoor use, but these have always included a ballasted base of such weight as to preclude its being blown over by the strongest natural storms. When the size was appropriate, horizontal bars, integral with the structure have been incorporated, which formed a ladder up the sloping back of the tower giving access to a platform about 1.2 m (4 ft) below the windshaft. Invariably, all timbers meant for outdoor use have been given preservative treatment.

Except when the windwheel is well above head height, the author reminds the reader that a test area, indoors or outdoors, should be surrounded by safety screens so that it is not possible for a person negligently to walk into danger.

A prototype machine should be of high mechanical excellence. During many of the tests the windpower available may be small and mechanical losses can then take on a relatively large proportion of the available power.

Bearings

Bearings are machine elements whose function it is to restrain a shaft laterally and endwise while leaving it free to rotate about its axis. Those responsible for lateral location are known as journal bearings. Those which locate a shaft endwise are called thrust bearings. The special case of a thrust bearing which takes the weight of a vertical shaft is called a footstep bearing.

A guiding principle (not absolutely universal) is that shaft and bearing should be of different materials, for like materials do not always work well together. A common practice in modern work is for steel to bear on brass or bronze. At low speeds hardwoods can make excellent bearings. These are generally made with the grain of the wood parallel to the axis of the shaft. (This was for a long time normal for the tail-end bearing of a ship's propeller shaft. The steel shaft was encased in bronze and the bearing was lignum vitae arranged in strips parallel to the shaft. A controlled inward leakage of water was allowed as a lubricant.)

Hollow shafts

The removal of the centre half of a solid shaft detracts but little from its ability to withstand bending or torsion. In applications where weight is important, such as motorcars and aircraft, tubular shafts where much more than the centre half is lacking, are commonly used. This does not mean that iron piping made for carrying fluids, or as a conduit for electric cables, is necessarily suitable for use as the windshaft of a mill. Some pipes are not of structural quality while others may be strong enough. In any case it must be remembered that the inside of a pipe might be the scene of hidden corrosion and that the outside is unlikely to be made to an exact size suitable for the use of standard bearings. Given sufficient knowledge and experience on the part of the builder a hollow shaft can give improved strength for a given weight of material.

Fastenings

Structures of stone and earth are chiefly held together by the force of gravity. Their intrinsic strength is compressive and some shear strength is supplied by friction between components and the adhesion of such mortar as may be used. Tension induced in such structures may be reduced by avoiding vertical walls or by the use of a batter (the inward leaning of walls and banks).

Extension of technology requires a material which can withstand tension. To some extent this can be supplied by stone, witness the "through" stones of dry-built walls, but stone is limited in this respect. Timber was an early tensile material. The Minoans used it as a reinforcement in their masonry, as protection against earthquake damage. Characteristically, wooden beams were trapped in place by the weight of the stones above. But they went farther than this. Most stone walls were built from the two faces with packing in between.

There is at Knossos evidence of timber (secured by dovetail ends) being used as ties between the two faces.

The wedge, the dovetail, the pinned mortise, the lap joint and the dowel open the way to a very wide technology. These are still basic in woodwork.

The dowel, or as it is called in structural work, the trenail, i.e. a form of wooden nail, is indeed often preferable today to its iron counterpart in that it is less likely to provoke corrosion and decay, and its larger size means that the load is spread over a bigger area and crushing stresses are diminished.

The majority of small windmills will probably be built from resources in between the examples quoted above. Indeed there are regions and times when scrapped motorcars are regarded as basic raw material. These are not always to be despised. The author has found that ex-automobile rear axle half-shafts, and brakes (particularly the older type of drum brake with cam-operated shoes) have been useful for windmill purposes.

Construction in primitive circumstances

The foregoing notes on construction have assumed that the specialised products of an advanced engineering society are readily at hand. Construction may be most needed, however, where very different circumstances prevail. Guidance may then come from study of how things were done in the past, particularly in countries poor in natural resources.

The windshaft

The object of a shaft is either to transmit mechanical power along its length or to support items of rotating machinery. In the case of a windmill the windshaft has to support the windwheel at one end and a power take-off, such as gearwheel or crank, at some point. Whilst it is natural in an advanced industrial context to consider that a shaft should be round, in primitive technologies it is often square and it may be formed either in wood or iron, locally worked, and only in circular cross-section at the bearings. A reason for this is the comparative ease with which wheels can be attached to and driven by a square shaft. Such wheels are made with a square hole in the centre. This slides loosely over the shaft. The wheel is then made to run true and is secured by means of wedges. The square shaft also lends itself to four windwheel arms being mortised in, one on each side—or to eight arms if the end is trimmed into an octagon.

Wooden machine elements can in general only be considered for low speed applications, that is, where any dynamic out-of-balance forces are small. In the traditional windmills of the North-West the neck bearing section of a wooden windshaft was normally protected from wear by metal strips which were set longitudinally into the wooden shaft and secured in place with iron bands. The tail bearing could be an iron gudgeon set into the end of the shaft. The thrust was often taken in this place on an iron collar. The author has also seen this type of construction in Portugal. There is no such protection for the wooden windshafts of the ancient mills of Crete, where wood bears on wood. Smaller shafts of wood, such as are used in the small vertical-shaft water mills of the Aegean, normally have an iron gudgeon in the lower end and a much longer iron insert at the top end, which also serves as the drive shaft to the millstone. This iron insert has a long flattened tang where it goes into the wooden shaft.

The square iron shafts of the irrigation mills of Crete are worked into a circular section at two points for the bearings. This is apparently done by forging rather than by turning. The bearings are of wood which has become soaked in oil.

Where a horizontal shaft always carries a vertical load the upper half of the bearing may not be necessary but a guard of some type, normally clear of the shaft, is an essential safeguard should the rotating shaft be accidentally forced upwards.

Note on the traditional mill

Fifty years ago an American writer (A.S.M.E. 1927–28) wrote rather severely that the traditional type of windmill was not distinguished either for efficiency or starting torque and, indeed, it is easy to criticise an apparently extravagant use of high-grade timber. However, the very mass of the windmill must have given an advantageous flywheel effect tending to uniformity of motion in a region of blustery winds. The characteristic form seems to have developed from radial arms pierced to carry transverse rods. Through these rods straw or reed was woven to form a kind of mat. Thus local materials were used. Old illustrations suggest that this form of sail was in general use from Crete to north-west Europe. Mediterranean development as the reader has already gathered was to a sail, as on a ship, flying freely so as to form its own aerodynamic surface. That of the North was to a rigid grid of timber over which canvas was tightly stretched. Over the years the grid moved more to the trailing side of the sail arm showing increasing appreciation of the importance of the leading edge. In Britain the canvas sheet largely gave way to hinged shutters which could be controlled while the mill was at work. Finally in the Netherlands came the shaped leading edge (associated with the name of Dekker) approaching that of the aerofoil.

11 Constructional materials: strengths and weaknesses

Even in normal working a windmill is subject to constant changes of stress, this in itself a severe condition, and the installation is of necessity in the lower and more turbulent part of the atmosphere where it is subject to constant buffeting and atmospheric corrosion. Every step should therefore be taken to choose appropriate materials. The following survey will provide some guidelines.

Whilst exceptionally brick, stone, or concrete may be used for a small windmill tower, mild steel or timber are more likely to be used. Aerodynamic surfaces may be of wood, canvas, or glass reinforced plastic. Soft grades of aluminium sheet formed over wooden supports have been employed but in some cases their use has had to be suspended because of their acting as disturbing radar reflectors!

Mild steel

Mild steel is the material produced when iron, which has been highly purified in the molten state, has a small, controlled quantity of carbon (between one tenth and one third of one per cent) added to it. This is made to enter into a definite chemical and physical relationship with the main mass. Mild steels include most of the structural steels of commerce. When of good quality they are one of the most reliable of all materials.

Aluminium and the light alloys

Although aluminium is very abundant in the crust of the earth it is so difficult to extract from its ores that it was not produced in the metallic form until early in the nineteenth century. It was then a scientific curiosity, with the status of a semi-precious metal. Economic production was able to begin after 1886 when the electrolytic process was invented. Subsequently, with the development of hydroelectric power, the metal has become abundant. In the commercially pure form it is soft and ductile but when scientifically alloyed and heat-treated, materials can be produced which attain the strength of steel at one third of the weight. However, its use when stresses are high needs specialist knowledge and laboratory control. The high strength so carefully developed can be lost by careless treatment and local corrosion can be induced by contact with incompatible metals.

Glass-reinforced plastics (G.R.P)

These are recent man-made materials of great versatility. Certain synthetic resins, which initially have the form of a viscous liquid, can be made to solidify when mixed with an appropriate chemical and then subjected to a carefully controlled curing process. Since the resulting product may lack tensile strength it is usually combined with a glass-fibre reinforcement. Glass fibres are one of the strongest materials known to man. Three stages are generally involved in the production of a glass fibre article. First is the making of the pattern or, as it is sometimes called, the plug. This may be of wood, or sometimes of cement smoothed with plaster of Paris. If the final product is an aerodynamic surface then the first pattern must have as perfect a surface as that intended for the final product. When the pattern is complete a mould is made around it (this itself may well be made of glass-reinforced plastic), the pattern being first coated with a release agent to prevent adhesion. The mould may have to be constructed in many separable parts so that it can be removed from its pattern and the final product released from the mould. Finally the articles to be made must be laid up in the mould, one at a time, and then subjected to the curing process. At its best G.R.P. can be very good, as is shown by the many thousands of satisfactory yachts which have been built during the last two decades. The excellence of the product, given perfect raw materials and design, depends still on the skill of the craftsman and the correctness of the curing. G.R.P. lends itself well to surfaces which have a double curvature and economy is helped when the cost of the plug and the mould can be spread over a large number of identical items.

Timber

At its best timber can produce a combination of strength, lightness, and stiffness which is unsurpassed by other materials.

There is a great range of natural timbers with an even greater range of properties. Durability depends much on the species and in some timbers there is a danger of local weakness due to knots and other discontinuities. Use for a significant structure needs knowledge of the species and a craftsman's eye in the selection of pieces to be used. Chemical impregnation is often an aid to durability. Iron bolts in some timbers can cause local deterioration.

Factor of safety and fatigue

A structure, in so far as the stresses in it can be determined, is designed so that under working conditions it is only loaded to a fraction, perhaps one tenth, of what is known to be the ultimate collapse load. The ratio of collapse load to the working load is called the "factor of safety". Its value depends on the engineer's estimate of certain factors which might arise.

In real life, only rarely are structures destroyed by a single overwhelming load. More often the process is insidious. This is called fatigue. Over a period of time, during repetitive loading, and when loads are coming on and off, a minute crack can be formed which gradually spreads through the main mass of material. Ultimately a sudden shock causes the crack to spread rapidly and fracture results. Fatigue is a well documented phenomenon in metals and suspected in some other materials. Fatigue failures came to the fore in the early days of the railways. Unfortunately the lesson of the liability of some materials to fail by fatigue seems to have to be learned anew in each generation. A fatigue fracture in a metal has a characteristic and easily recognised appearance. Formerly it was sometimes believed that the metal had changed during the fatigue process. In fact tests adjacent to a fatigue crack often show that the metal has not changed at all. Mild steel has a fortunate property. If the stress at which failure occurs is plotted against the number of stress cycles to failure, then a stress-endurance curve results (Fig. 11.1). At a certain stress this curve flattens out, indicating that at any lower stress than this the material can go on indefinitely. The stress at which the curve flattens out is called the fatigue limit. At stresses above this limit the material has a finite life. Below it the material can apparently go on for an indefinite time.

The fatigue limit is not a universal property. In some materials the endurance curve continues to fall (Fig. 11.1(B)). It then becomes a

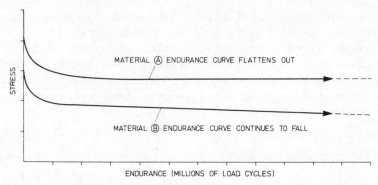

Fig. 11.1   Metal fatigue: endurance curves

matter of sufficient stress applications for failure to occur. For certain structures which use high-strength light-alloys, failure is not so much a probability as an inevitable destiny if the structure remains in service long enough (measured in stress cycles). This leads to the concept of limited life for each component of the structure. Laboratory tests suggest a safe life. At the expiration of this the structure must be resolutely scrapped even though no damage is apparent on the closest examination.

Corrosive influences can increase the danger of fatigue failure. Minute corrosive pits form foci for the initiation of local damage and ultimate failure by fatigue. Marine atmospheres can be particularly vicious. The failure of stainless steel yacht fittings is by no means unknown. The many attractions of aluminium and aluminium-alloys must be balanced against their lack of a definite fatigue limit, bearing in mind that buffeting and vibration may continue even when a machine is at rest.

12   Miscellaneous

Windmills afloat

Things so common in daily life as not to be mentioned in normal
records can, in the course of a few years, be utterly forgotten.
Sometimes they may be recalled by a passing reference in a book on
quite a different subject. Such was the case when the author was
reading "Falcon in the Baltic" by E. F. Knight. Knight was a Victorian author and journalist and an almost incredible yachtsman. At
one stage in his career he bought a ship's lifeboat which he named
"Falcon" after converting her to a cruising yacht. In 1887 he set off
for the Baltic but when once at sea "Falcon" began to leak. (This was
ultimately tracked down to a keel bolt which had been omitted in the
conversion, the hole having been blocked by Thames mud until the
sea washed it out.) Approaching the Dutch coast (and pumping hard)
Knight engaged a pilot to take him into Voorne. The pilot was
evidently critical and Knight was sensitive. Subsequently the pilot
tried to make amends and the following dialogue is reported.

"Capital boat to run before a sea, this, Captain," he said after a
few minutes, "but look, there is Goeree." As he spoke the rain had
ceased, the sky had cleared a bit, and there, before us about a mile
away to leeward, suddenly appeared a low, pale green shore with
several hurrying windmills in the background. . . . Across the dykes
of Voorne we perceived an enormous congregation of windmills.
"What do you have all these windmills for?" asked Wright (the
paid hand). "To pump de water off de land," replied the pilot. "If
those was not always turning round us Hollanders would soon all
be drowned." "Well pilot," I said, "you were very severe just now
about our boat's leaking, but you must confess that your country
leaks harder still. Your windmills are always pumping just as they
do on an old Norwegian timber vessel." He chuckled softly and
replied merely, "I think, Captain, I will take one little drop more
of that rum."

This is one of the definite historical records of the fact that
Norwegian vessels engaged in the timber trade between Britain and
the Baltic did characteristically mount windmill pumps to drain the
bilge. Subsequent reading showed that the ships were probably built
of softwoods and when they were old they were engaged in a highly
competitive trade. Not to pump would have been to sink (but their
timber cargoes might keep them just awash). Hand pumping would
have imposed too big a demand on the crew. Hence the windmill.
Underhill records that these ships were commonly known as
"onkers", the name probably arising from the characteristic sound of
the windmill pump. Its noise would spread across the water and be
easily picked up by other sailing vessels which, having no engines,
would have little background noise to hide it. "Onkers on the port
bow" may well have been the lookout's cry on a misty night.

A very famous Norwegian ship of this period rigged a windmill for
electric light. Polar explorer, Nansen, allowed his ship "Fram" to
become locked in the Arctic ice, hoping that the drift would carry
him near the Pole. On the 25th October, 1893, the windmill was
started up and Nansen records how the spirits of the ship's company
rose as the lights came on. This windmill worked for about one year
but he records that by June 1895 the cog wheels had worn out.
Illustrations show that this machine had four blades and, by scaling
against the mast, it looks about 6.3 m (20 ft) in diameter. Nansen also
shipped a horse-mill, modified for hand working, which he planned
to use should there be absence of wind. His colleagues needing exer-
cise were to supply light to those with work to do below! In the event,
all the ship's company always had work to do below and mercifully
that machine was never used.

The day of the seagoing windmill has come again. Formerly
coastwise navigation was based on the chart, the compass, the
mechanical log and the lead line. Now depths, speeds, distance run,
wind speed and direction (both absolute and relative to the yacht) are
all read off electrical instruments. Furthermore checks in position
which formerly depended on the sighting of navigation marks can now
be made by the bearing of radio beacons. All these devices can be
powered by a battery which must be kept in a state of charge. The
auxiliary engine can do this if there is one and a little windmill can be
used when the yacht is swinging on her moorings. The power
required is very small and so the windmill, having the strength of
small things, can be simple and stormproof. Such windmills are

available commercially and can now be seen at work in most yacht harbours.

The suggestion that ship-mounted windmills might drive paddles or propellers which in turn push the ship along is an old one but one which constantly comes up again.[15] The attraction is a hoped-for ability to set a course without reference to the direction of the wind, if need be, straight into the wind's eye. An unpublished analysis by Professor Preston suggests that it might be achieved subject to a certain size ratio between windwheel and propeller and to very perfectly designed components. Some such devices have been built and they have claimed a degree of success but the author has not been able to find any records of systematic and comprehensive testing. The assumption must be that their success was so marginal as to outweigh their practicability.

The high-speed, free-spinning windwheel

A narrow bladed, free-spinning windwheel, generally of a symmetrical blade section, and mounted without pitch angle, can rotate at very high speed and in doing so act as an almost total obstruction to airflow through its disk. As mentioned previously, such have been used as windscreens.

One application of the free-spinning windwheel, with almost vertical axis, was the autogiro, an aircraft which was an approach to hovering flight without stall.[16] Another application, associated with the name of Moore-Brabazon, was its use instead of a sail on a yacht. The boat was a "Solent Redwing", 6.7 m (22 ft) in length and the rotor, mounted on an alloy mast, was 5.4 m (18 ft) in diameter. The maximum rotational speed was 250 rpm, with a tip-speed of nearly 72 m/sec (160 mph). A new sailing technique had to be developed. Moore-Brabazon considered that the performance was very remarkable both as regards speed and the ability to point high into the wind. However, before it could be put to test against orthodox sail craft it had drifted on to a dinghy—with results which were disastrous. Moore-Brabazon concluded that as a cruising rig the rotor is dangerous and impractical.[16]

The mounting of a windmill on a barge or pontoon so as to provide a negligible friction turntable is reported from the Netherlands, as an isolated example, applied to a saw mill. The author considers that the method has apparent advantages which could be employed in a place where water is abundant, and land use is pressing.

The storage of energy

The problem of harmonising supply and demand is met in personal incomes by the services of a bank. Incomes are deposited, to be withdrawn as required. The warehouse is an analagous case which smooths out the flow of commodities. A lot of windmill problems could be solved if we had an energy bank. Many ideas can be discussed and some can be of limited use, but a wholly satisfactory energy store does not yet exist.

The heat store

A simple and widely available energy bank is a heat store, such as a night-store heater or insulated water tank. This is convenient and cheap but while it accepts the highest grades of energy its repayments are utterly degraded. To use a financial analogy, whilst it accepts deposits in an international currency it repays with truck tokens. The highest form of energy, electricity, is fed in, and this could otherwise have done more useful things such as working an electric drill or driving a radio set. The repayment is in such a low grade that it can only be used for domestic purposes—such things as washing your feet or filling a hot water bottle!

Water is widely available, generally cheap and non-poisonous and non-corrosive. Also it has a large energy capacity. It accepts a lot of energy with only a small rise in temperature. It would be possible (but hardly practical) to recover a little of the heat as high grade energy if we made our storage tank in the form of a steam boiler and by increasing the pressure got the water temperature well above atmospheric boiling point. Locomotive engines for use in hazardous areas were formerly built without a firebox. The "boiler" was filled with water at high pressure and temperature and periodically, as duty required, was recharged with steam. As the pressure was released some of the water flashed back into steam and this was used to work the engine. The range was limited.

The cold store

The reverse process, of storing cold instead of heat, should be entirely satisfactory. A refrigerator or deep-freeze can have a lot of inherent storage to carry over periods of calm. Suitably designed plants could become heat pumps with the property of discharging far more heat energy than the mechanical energy which they received from the windmill. The process of cooling the larder can be combined with heating the domestic water. This idea is sometimes startling to people without an engineering background but it is completely sound and does not contravene any natural laws. It is a very old idea which comes up in times of energy stress and is then quietly forgotten when fuel prices fall.

The lifted weight

Storing energy in such a way that it can be recovered in a high grade form is difficult to carry out on a practical scale. One method is by elevation. A weight lifted against gravity has work done on it which can be given back when the weight is allowed to fall. The weights of a grandfather clock store enough energy to drive the timepiece for a week, and the water which flows into a miller's pond stores enough energy during the night to drive the mill during the day. A windmill could of course pump the tailwater back to the head pond and then your problem would be solved, provided you had a mill pond. This is a very old and a very good solution but not always a very practical one in a crowded land. It is of course used on a very large scale to aid the economical running of the national electricity grid. It was certainly proposed in connection with wind pumping in Australia about eighty years ago and it could well have been used before that. Watt's first steam engine was used to back-pump a waterwheel plant. It would be surprising indeed if a windmill had never done the same, and they have certainly been used for canal pumping. On a domestic scale, if you pumped water up to a tank on the roof, the amount of energy stored might be disappointing. A tank as big as a living room, at roof level in a two storey house, would store enough energy to run a food mixer for about twenty minutes. For small and precise applications such as a dentist's drill, an instrument maker's lathe, or even a potter's wheel it might be just about adequate.

Flywheel storage

Another way of storing mechanical energy is to set a mass of material into motion, a battering ram in linear motion, or a flywheel in rotary motion. Nearly all reciprocating engines store up energy of motion so that the intermittent forces on the piston can appear as a steady torque on the output shaft. The amount of energy which can be stored in a flywheel is large but it has disadvantages for other than a short period. Flywheel bearings and air friction are a constant drain of energy. The proper application of flywheel stored energy is to provide a very large power over a very short time. Such duties as piercing or cutting through slabs of steel are appropriate, as are such devices as have enabled one little man, over a period of a few minutes, to store up enough energy to start a heavy diesel or aeroplane piston-engine. There is however a lot to be said for a moderate flywheel in a simple workshop application of wind power. It could make the speed more uniform by reducing overspeed in gusts and a slowing down in lulls. An electrical generator, however, generally has to run so fast that its rotor has quite a lot of flywheel effect.

Storage by elastic deformation

The third method of storing mechanical energy is by the elastic distortion of a solid substance or by the compression of a gaseous one. A special but rather small case of the first is a clock spring which is wound up once a week and gives back its energy in between; or of a bow, which stores energy as the string is drawn back, to return it rapidly to the arrow when the string is released. Another example is an air gun. Energy is first concentrated in a steel spring. This is released to compress air until the projectile begins to move. The air then expands, doing work behind the projectile. A compressed air reservoir could be charged and hold its contents without loss for a long period of time. But on the whole it is better suited to the meeting of peak loads than to steady output over a long period of time. A compressed-air receiver is also a completely unacceptable explosion hazard unless it is under engineering supervision.

Electrochemical storage

The passage of an electric current through certain substances can lead to chemical changes which are reversible, in that given suitable conditions the chemical substances will go back to their original state and in doing so restore some of the energy, as electricity, that was first given to them. This gives the virtual impression that energy is being stored and it is the basis of the so-called accumulator battery, a device well known as an adjunct to the motor car. Failing the possession of a mill-pond, electrical accumulators can be regarded as the only really practical means of medium term small-scale energy storage. The best known examples are based on the inter-action of lead compounds and sulphuric acid. These are followed by those which use nickel and caustic potash. Both are widely and successfully used. They are, however, products of "high technology", are expensive, and contain unpleasant and dangerous chemicals.

Another electro-chemical method is that of the electrolysis of water, whereby this is separated into its constituent gases, oxygen and hydrogen, the latter being used as a fuel, after any period of storage. Not a new idea—it figures in a Danish Government programme of over 74 years ago—it is still far from adequately developed and while mentioned here for its interest and possibilities, it has serious inherent dangers and the author strongly discourages the reader from experimenting. (The recombination of the two gases can take place with great violence—truly explosively—and accidents can happen. . . . The warning found on some accumulators "Battery gases can explode" is serious and appropriate—as the author will not readily forget from such an experience with an overcharged battery.)

Electricity grid

A "reservoir" which could always receive small amounts of energy is of course the national electricity grid. This is capable of absorbing wind energy as and when it comes. Several traditional windmills in the Netherlands have had dynamos (induction generators) fitted to enable them to do this. If bits of wind energy could be fed into a system exactly when they were needed the grid authorities might welcome them as an aid to peak load and to regulation problems, even though they might not wish to pay enough to make it a tempting commercial investment.

REFERENCES

1. *Cornish windmills*. H. L. Douch. Blackford Ltd, Truro, c1960
2. Economic windpower. D. E. Elliott. *Applied Energy*, I, 1963, 167–197
3. Wind power in Eastern Crete. N. G. Calvert. *Trans. Newcomen Soc., XLIV, 1971–2*
4. Windmill sails. R. Wailes *et al., Trans. Newcomen Soc.*, **XXIV**, 1945, 150
5. *The mills of Lasithi*. N.G. Calvert. Agios Nicholaos. Almathea, 1973
6. The monokairos windmills of Lasithi. N. G. Calvert. British School of Archaeology in Athens. *Annual*. Vol **LXX**, 1975, 51–57
7. The characteristics of a sail mill. N. G. Calvert. *Jnl Industrial Aerodynamics*, **3**, 1978
8. Water mills in Central Crete. N. G. Calvert. *Trans. Newcomen Soc.*, **XLV**, 1972–3
9. Horizontal windmills. R. Wailes. *Trans. Newcomen Soc.*, **XL**, 1967–8
10. *The story of the Rotor*. A. Flettner. Willhoft, New York, 1926
11. *The wing-rotor in theory and practice*. S. J. Savonius. Helsinki, 1926
12. Who put the speeds in Admiral Beaufort's Force Scale? Blair Kinsman. U.S. Inst. Navigation. *Oceans*, 1958, 18–25
13. Report on the use of windmills for the generation of electricity. Inst. Research in Agric. Eng., Oxford. *Bulletin No. 1*. 1926, and *Abstract*, 1933
14. Wind energy in the United Kingdom. R. Rayment. Building Research Establishment, Garston. *Current Paper 59*, 1976
15. *Windmill boat references:*
    (i) A floating engine invented by the French for the invasion of England. (A floating castle mounting three windmills and four paddle wheels. Intended to carry 60,000 men and 600 cannon). Date unknown. *International Symposium on Molinology* (Denmark Meeting), 1969
    (ii) A ship mounting a sail windmill connected to paddle wheels. Attributed to Duquet, 1712. (Appears in (10) above.)
    (iii) A similar case to (ii) attributed to Smit, Holland, 1890. *Scientific American*. New York, 1925
    (iv) A propeller-driven launch powered by a 9 m (30 ft) aerofoil windmill. Constantin, France, 1924. (Appears in (10) above.)
    (v) A light propeller-driven catamaran powered by 4.25 m (14 ft) windwheels. *Jnl Yacht Research Assoc.*, Hudson, Ohio, 1966
16. The Autogyro rotor as a sail. J. Moore-Brabazon. *Jnl Royal Aero. Soc.*, Vol **XXXVIII**, 1934, 771

SUGGESTED FURTHER READING

The generation of electricity by windpower. E. W. Golding. Spon, London, 1976
Power from the wind. P. C, Putnam. Van Nostrand-Reinhold, New York, 1974
Fluid mechanics: a laboratory course. M. A. Plint and L. Böswirth. Griffin, London, 1978

HISTORICAL
The English windmill. R. Wailes. Routledge, London, 1967
Windmills and millwrighting. S. Freese. David & Charles, Newton Abbott, 1971
The Dutch windmill. F. Stokhuyzen. Merlin Press, London, 1962
Wind energy bibliography. Windworks. Mukwanago, Wis., 1973

Index

Netherlands, 9, 10, 70
Newton, Sir Isaac, 19, 23
Noise, 29, 51

"Onkers", 112
Oscillation, torsional, 40
Overspeed, 50

Paltrok, 33
Pilot plant, 99
Pitch-angle, 31, 51
Pitot-static tube, 82
Plastics, glass-reinforced, 108
Post mill, 10
Power measurement, 86, 95
Prandtl, Ludwig, 64
Pressure wave, 29
Preston, Prof. J. H., 38
Pumping, 10, 36

Radio, powering, 11, 112
Regulation, 55
Relative velocities, 26
Reversibility, energy, 17
Reynolds, Osborne, 19, 20
Rigging, sail-mill, 57
Rotor ship, 64
Runaway, 50

Sail, 54, 57, 58
Sail, aerofoil comparison, 57
    rigging of, 57
Savonius, S.I., and rotor, 66
Separation, 21
Shafts, 103
Shape, blade, 39
Similarity, 93
Site, 96
Slip rings, 34
"Solidity", windwheel, 31
Stall, 24, 75

Static friction, 18
Streamlines, 22, 35
Speed-tip ratio, 42
Supports, 31
Sweeps, 39
Symmetrical aerofoil, 57

Tail vane, 34, 77
Testing, 84, 88, 90
Tip-speed of blade, limit of, 51
    ratio to wind speed, 42
Tjasker mill, 36
Torque, 36, 86
Tower, 31
Transition, 21, 22
Transmission, electric power, 34
Turbulence, 18, 20
Twist, (helix) of blade, 53

Units, of measurement, 85

Velocity diagram, 26
Vertical axis, 59
    antiquity of concept, 60
    Fairbairn's stricture, 61
    Musgrove, P.J., 71
    South Rangi, 61
Viscosity, 20, 21

Windpower, at sea, 112
Windmills, aircraft, 11
    American, 10
    Cretan, 14
    industrial, 10
Windseeking, 32, 77
Windshaft, 32
Windwheel, 30
Wip mill, 33

Zuider Zee, 10